Calculation of Roundabouts

Raffaele Mauro

Calculation of Roundabouts

Capacity, Waiting Phenomena and Reliability

Springer

Prof. Raffaele Mauro
Università Trento
Dip.to Ingegneria Meccanica e
Strutturale (DIMS)
Via Mesiano, 77
38050 Trento
Italy
raffaele.mauro@ing.unitn.it

ISBN 978-3-642-04550-9 e-ISBN 978-3-642-04551-6
DOI 10.1007/978-3-642-04551-6
Springer Heidelberg Dordrecht London New York

Library of Congress Control Number: 2009939930

Cover design: eStudio Calamar S.L.

Printed on acid-free paper

Springer is part of Springer Science+Business Media (www.springer.com)

To my Mother, Invigilans Lucerna

Preface

Over the last decades, roundabouts have been increasingly used in building new at-grade intersections and in changing the layout of existing intersections. Therefore, we decided that it would be useful to collect, in one volume, the methods and procedures used to evaluate the operating conditions of this type of intersection.

The organization of this book is as follows:

Chapter 1 deals with the definitions of capacity, capacity indices for roundabouts, and parameters linked to waiting phenomena at entries, i.e., delays and queue lengths.

This is preceded by an introduction to the fundamental concepts associated with statistical equilibrium and steady-state conditions. These general concepts of the theory of systems and control are applied to roundabouts.

Chapter 2 starts with some examples of capacity formulas, selected from the three types that are available in today's scientific and technical literature. Then, some criteria for taking into account the effects of pedestrian flows on entry and exit capacities are presented. The remaining part of this chapter is dedicated to the calculation procedures for flows entering the roundabout, to the capacity in case of saturation or oversaturation of one or more entries, and to simple capacity and total capacity.

Chapter 3 covers the analysis of waiting phenomena under steady-state and transient conditions.

The material in Chap. 3 includes the standard key results of simple probabilistic (Markovian) and deterministic queuing systems, as well as the results of some special time-dependent solutions for waiting phenomena.

Chapter 3 has a more general scope because it presents results that may be used both for roundabouts and for any at-grade intersection.

Chapter 4 deals with the application of the results and methods discussed in the previous chapters for the evaluation of waiting times, queue lengths, and levels of service of roundabouts. The calculation procedures illustrated are meant for operating conditions characterized by undersaturated, saturated, or over-saturated entries.

Chapter 4 ends with the determination of the level of service of a roundabout.

All the chapters include worked examples that have been developed in sufficient detail to explain as clearly as possible the formulas and the procedures presented in the book.

We do hope that the expository format that we have used, which is characterized by a plain style and supported by worked examples, will help the reader to easily understand and be able to use the materials discussed in the book. We also hope that this layout will help the highway traffic engineers analyze the operating conditions of roundabouts.

Chapter 5 presents criteria to evaluate roundabout performance reliability. After introducing and justifying the adoption of reserve of capacity and rate of capacity as performance functions, the discussion is developed using a general calculation criterion in which the values that are involved in the limit state service condition – traffic demand and entry capacity – are random variables described by their probability density functions, that is to say by their distribution functions.

A lower level criterion is then identified with which, on the basis of the estimation of suitable statistics of the performance function, a reliability index is calculated that can be compared to a prefixed reference value.

Trento, Italy *Raffaele Mauro*

Acknowledgments

My continuing interest in the study of roundabouts is due to my mentor at University of Naples "Federico II", professor Tommaso Esposito, who, over ten years ago, urged me to publish the first two works that appeared in Italy on this type of intersection. Most of this book is the product of our common studies and consulting work.

I have often discussed with Giulio Erberto Cantarella, professor of transportation system theory at Salerno University, about steady-state conditions, transient states, and time-dependent solutions, and he gave me many fruitful suggestions and helped me focus my attention on interesting and important aspects of the topic.

Special thanks also to Michele Corradini and Alessandro Romania. Michele Corradini and Alessandro Romania, both of whom have M.S. degrees in Civil Engineering and who were my students at Trento University, designed figures, checked analytical developments, verified the accuracy of numerical calculations, and offered numerous, relevant observations.

This edition is based on an English translation prepared by Dr. Savino Carrella. He also has patiently revised, corrected, and improved the manuscript.

Finally, I wish to thank the Province of Trento that has supported with a grant my recent researches on roundabouts (Special Project CRS – Performance analysis of roundabouts).

Contents

Chapter 1
Calculation of Roundabouts: Problem Definition

By analyzing roundabouts, we mean the determination of efficiency measures when these intersections are to be used by a known traffic demand.

The indices that we generally take into account are:

Capacity and other related indicators
Queue lengths
Waiting times and delays

In traffic engineering, queue lengths, waiting times, and delay values are also indicated as measures of effectiveness (MOE). Capacity determinations for entries refer to capacity, capacity rate, and reserve capacity, which is expressed in both absolute terms and in percentages. Capacity evaluations for the whole roundabout refer to simple capacity, total capacity, the mean of reserve capacity, and the mean of capacity rate at entries.

Queues result from the waiting phenomena that drivers may suffer at the entries, and we estimate the length of the queues, measured by the number of vehicles, in terms of means and percentiles.

Waiting times are the result of queuing up. They increase the traveling time because of the intersections along the route.

Waiting times may refer to single users, as value or expected value[1], or, with appropriate means, to the whole intersection.

When we calculate capacity, queue lengths, and waiting times, we must specify, for the observation period chosen, steadiness and variability of traffic demand and the presence of one or more entries that are saturated or oversaturated.

This involves the analysis of the intersection with and without statistical equilibrium and, according to the state of the entries, the use of probabilistic, deterministic, or time-dependent models.

---

[1]Expected value generally means the average of a random variable. The terms mean, average, and mathematical expectation (for a random variable) are synonymous.

R. Mauro, *Calculation of Roundabouts*, DOI 10.1007/978-3-642-04551-6_1,

In the remaining part of this chapter, we will specify the assumptions and the terms for calculating roundabouts, and we will also give specific definitions of the performance indices that are briefly listed above.

In the following chapters some of the procedures used to calculate roundabouts will be illustrated.

1.1 Statistical Equilibrium and Steady-State Conditions

The operational conditions in the time of a roundabout – as is the case for any uncertain, real system – may be considered as a succession of states with probabilistic characterization.

The description of this evolution requires, in its most complete forms, knowledge of the probability associated with each state of the system.

This probability, for the same state may vary any time. In this case, we say that the system exists in a transient condition.

If the probabilities of the states remain constant with time, we may say that the system has reached a statistical equilibrium, and we denote it to be in a steady-state condition.

It is now worth noting that, to evaluate if a system is at the steady-state condition, we often choose not to assess the time-invariant state probability distribution.

Thus, the assessment of a steady-state condition is only about the constancy with time of appropriate statistical values (e.g., means, variances, joint moments, and r-order moments, etc.) associated with one or more variables that evolve randomly and that are believed to be linked to the operational conditions of the whole.[2]

As we will see more clearly in the following chapters, on the basis of the above-mentioned second procedure used to confirm a steady-state condition, a roundabout is deemed to be at a steady-state condition when entering traffic demand does not vary with time, and, additionally, when the traffic is systematically served by the roundabout without the occurrence of the incoming traffic congestion phenomenon i.e., average queue lengths and waiting times that are indefinitely increasing with time.

We assume that all of the analyses performed below related to the operating conditions of roundabouts are based on the assumption that the circulatory roadway and the exits are always undersaturated. Therefore, the possible entry congestion must be related to one or more saturated or oversaturated entries.[3]

Now, if $i = 1; 2; ...; n$ is the number of the intersection legs and t is time, traffic demand at the entries is generally expressed [2] by a vector (vector of demand volumes)

---

[2]The two meanings correspond, in the random process theory, to steady-state conditions in restricted sense (strong steady-state conditions) and in broad sense (weak steady-state conditions), respectively. See, for example, [1].

[3]Evidently, it is assumed in this way that the absence of entry vehicular congestion is related to undersaturated entries.

$$Q_e(t) = [Q_{ei}(t)] \quad i = 1;\ 2;\ ...;\ n \tag{1.1}$$

and by a matrix of traffic percentages (percentage origin/destination matrix)

$$P_{O/D}(t) = [P_{ij}(t)] \quad i,\ j = 1;\ 2;\ ...;\ n \tag{1.2}$$

The elements of the origin/destination matrix of traffic demand $M_{O/D}(t)$ from entries "i" to exits "j" of the intersection

$$M_{O/D}(t) = [Q_{ij}(t)] \quad i,\ j = 1;\ 2;\ ...;\ n \tag{1.3}$$

is obtained multiplying each element of the row "i" of matrix (1.2) by the corresponding element $Q_{ei}(t)$ of vector (1.1). (See example in Sect. 1.1.1).

That being stated, the condition for having a steady-state system, as previously formulated, equals to the set of the following relationships

$$Q_e(t + \Delta t) = Q_e(t) \tag{1.4}$$

$$P_{O/D}(t + \Delta t) = P_{O/D}(t) \tag{1.5}$$

for any Δt and

$$Q_{ei}(t) < C_i(t) \quad i = 1;\ 2;\ ...;\ n \tag{1.6}$$

for any "t".

In Eq. (1.6) (entry undersaturation condition), $C_i(t)$ represents entry capacities, as defined in the following Sect. 1.2.

If Eqs. (1.4) and (1.5) are valid, we evidently also have

$$M_{O/D}(t + \Delta t) = M_{O/D}(t) \tag{1.7}$$

for any Δt.

In current technical practice, a steady-state condition is considered to be achieved, with undersaturated entries, if traffic demand at the intersection is constant for a finite period of time T, but long enough[4] to allow the stabilization of the operative conditions of the roundabout around the constant mean values E [·] of state variables.

In addition, the punctual values of state variables must be little dispersed around the mean values E [·].

It is worth recalling that we have used here the queue lengths and waiting times defined in the following paragraph Sect. 1.3 as state variables.

When Eqs. (1.4) and (1.5) are valid, and considering the preservation of flows at the intersection, for the vector of the traffic flow in the circulatory roadway,

---

[4]For the duration of T (in s) we generally use the following indication (Morse), $T > \max\{1/(\sqrt{C_i/3600} - \sqrt{Q_{ei}/3600})^2\}$, with Q_{ei} and C_i expressed in hourly volumes.

$$Q_c(t) = [Q_{ci}(t)] \quad i = 1; 2; ...; n \tag{1.8}$$

and for the vector of exiting traffic flows at legs

$$Q_u(t) = Q_{ui}(t)] \quad i = 1; 2; ...; n \tag{1.9}$$

we have

$$Q_c(t + \Delta t) = Q_c(t) \tag{1.10}$$

$$Q_u(t + \Delta t) = Q_u(t) \tag{1.11}$$

for any Δt.

In the worked example shown in Sect. 1.1.1, we explain how to deduce the vectors $Q_c(t)$ and $Q_u(t)$ starting with traffic demand at the intersection $\{ Q_e(t); P_{O/D}(t)\}$.

In technical practice, as is well known, the geometrical and functional design of a roundabout (or of any other infrastructural element) is generally conducted by assuming that traffic demand is constant with time and with reference to the traffic volume (peak-hour volume, PHV) related to a suitably chosen hour (such as, for example, between the 30th and the 100th peak hour of the year) [2].

For a generic entry, to obtain the design traffic volume, the peak-hour volume is divided by the peak hour factor, phf, to take into account traffic variations within the peak hour.

Working in this way, the design traffic volume is equal to the equivalent hourly traffic volume of the peak subhourly rate of flow [2].

However, strictly speaking, this procedure is not correct.

In fact, using the design traffic volume (obtained from the peak-hour traffic demand) which is considered to be applied indefinitely (in accordance with steady-state conditions), we have queue lengths, waiting times, and delays that are considerably greater than the ones actually applied.

In addition, the values of these parameters tend to become infinite under critical conditions, that is when demand nears, equals, or exceeds the entry capacity (saturated or oversaturated entry).

These circumstances cause an overdimensioning of the geometrical elements of the roundabout and an unrealistic evaluation of the Level of Service.

In reality, traffic peaks and critical conditions occur for more or less long time intervals, so that they are consequently finished with limited effects.

It is then suitable to base the evaluation on the demand flow trend inside the time period considered and, in addition, on the effective duration of traffic peaks.

Thus, as we will see in Chap. 3, the transient conditions of the system must be thoroughly analyzed.

1.1.1 A Worked Example on Traffic Demand at a Roundabout

Consider the roundabout in Fig. 1.1 and a steady-state condition in which traffic demand at the intersection, indicated by Eqs. (1.1) and (1.2), is the following

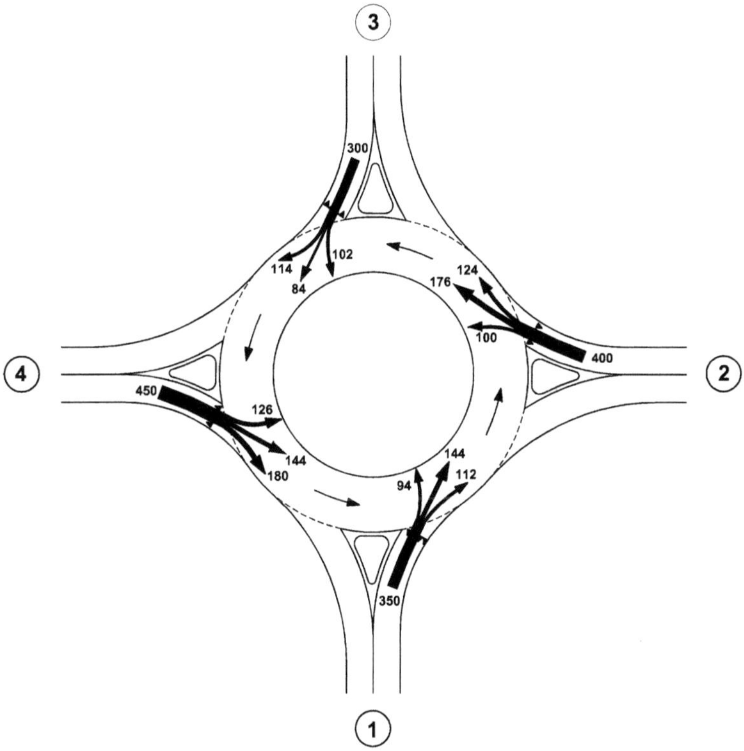

Fig. 1.1 Traffic demand at roundabout (volumes in veh/h)

(volumes are expressed in veh/h)

$$Q_e \equiv \begin{bmatrix} Q_{e1} & Q_{e2} & Q_{e3} & Q_{e4} \end{bmatrix} \equiv \begin{bmatrix} 350 & 400 & 300 & 450 \end{bmatrix}$$

$$P_{O/D} \equiv \begin{bmatrix} P_{11} & P_{12} & P_{13} & P_{14} \\ P_{21} & P_{22} & P_{23} & P_{24} \\ P_{31} & P_{32} & P_{33} & P_{34} \\ P_{41} & P_{42} & P_{43} & P_{44} \end{bmatrix} \equiv \begin{bmatrix} 0 & 0.32 & 0.41 & 0.27 \\ 0.25 & 0 & 0.31 & 0.44 \\ 0.28 & 0.34 & 0 & 0.38 \\ 0.40 & 0.32 & 0.28 & 0 \end{bmatrix}$$

Multiplying each element in row "i" of the matrix $P_{O/D}$ by the corresponding Q_{ei} of vector Q_e, we have the origin/destination matrix $M_{O/D}$ for the roundabout. (The terms are approximated to integer values.)

$$M_{O/D} \equiv \begin{bmatrix} Q_{11} & Q_{12} & Q_{13} & Q_{14} \\ Q_{21} & Q_{22} & Q_{23} & Q_{24} \\ Q_{31} & Q_{32} & Q_{33} & Q_{34} \\ Q_{41} & Q_{42} & Q_{43} & Q_{44} \end{bmatrix} \equiv \begin{bmatrix} 0 & 112 & 144 & 94 \\ 100 & 0 & 124 & 176 \\ 84 & 102 & 0 & 114 \\ 180 & 144 & 126 & 0 \end{bmatrix}$$

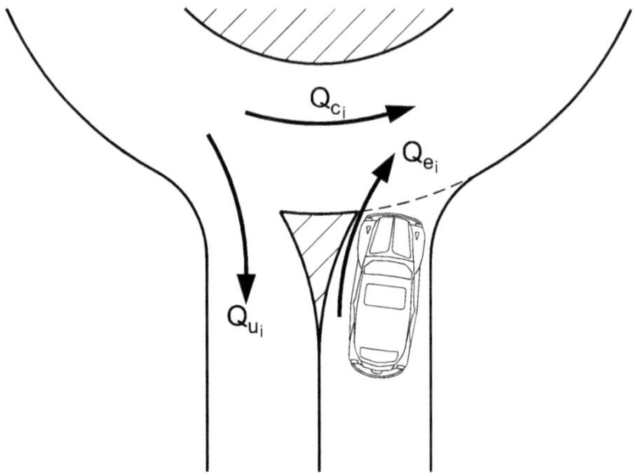

Fig. 1.2 Traffic flows at the roundabout for the entry "i"

From matrix $M_{O/D}$, due to traffic preservation on the circulatory roadway, it is possible to deduce (See Fig. 1.2) the volumes Q_{ci} in front of the entries and the volumes Q_{ui} exiting the roundabout with the following straightforward relationships:

$$\begin{aligned}
Q_{c1} &= Q_{42} + Q_{43} + Q_{32} = 144 + 126 + 102 = 372\,\text{veh/h} \\
Q_{c2} &= Q_{13} + Q_{14} + Q_{43} = 144 + 94 + 126 = 364\,\text{veh/h} \\
Q_{c3} &= Q_{24} + Q_{21} + Q_{14} = 176 + 100 + 94 = 370\,\text{veh/h} \\
Q_{c4} &= Q_{31} + Q_{32} + Q_{21} = 84 + 102 + 100 = 286\,\text{veh/h} \\
Q_{u1} &= Q_{21} + Q_{31} + Q_{41} = 100 + 84 + 180 = 364\,\text{veh/h} \\
Q_{u2} &= Q_{12} + Q_{32} + Q_{42} = 112 + 102 + 144 = 358\,\text{veh/h} \\
Q_{u3} &= Q_{13} + Q_{23} + Q_{43} = 144 + 124 + 126 = 394\,\text{veh/h} \\
Q_{u4} &= Q_{14} + Q_{24} + Q_{34} = 94 + 176 + 114 = 384\,\text{veh/h}
\end{aligned}$$

In the end, vectors (1.8) and (1.9) are, respectively:

$$Q_c \equiv \begin{bmatrix} 372 & 364 & 370 & 286 \end{bmatrix}$$

$$Q_u \equiv \begin{bmatrix} 364 & 358 & 394 & 384 \end{bmatrix}$$

Since the example is conducted assuming the system to be in a steady-state condition, and, therefore, traffic demand is constant with time and the entries are undersaturated, vectors indicating the different ways to express demand coincide with the flow vectors that are actually traveling through the intersection: in this case, therefore, Q_e indicates the traffic entering the roundabout, Q_c indicates circulating traffic in front of the entries, and Q_u indicates the exiting traffic.

However, from now on, corresponding volumes of demand and the flows in transit will be indicated with the same symbol; the meaning of the symbol, when it is not explicitly indicated, can be easily inferred from the context.

Evidently, calculation of the flows Q_{ci} and Q_{ui} may be conducted without the determination of matrix $M_{O/D}$ by using the elements of vector Q_e and matrix $P_{O/D}$ directly.

Therefore, due to the preservation of flows in the circulatory roadway and assuming, for the sake of simplicity, the flows that have the same leg as origin and destination to be null since they are generally very small, we have the following relationships:

$$\begin{aligned} Q_{c1} &= Q_{42} + Q_{43} + Q_{32} = (P_{42} + P_{43})\, Q_{e4} + P_{32}\, Q_{e3} \\ Q_{c2} &= Q_{13} + Q_{14} + Q_{43} = (P_{13} + P_{14})\, Q_{e1} + P_{43}\, Q_{e4} \\ Q_{c3} &= Q_{24} + Q_{21} + Q_{14} = (P_{24} + P_{21})\, Q_{e2} + P_{14}\, Q_{e1} \\ Q_{c4} &= Q_{31} + Q_{32} + Q_{21} = (P_{31} + P_{32})\, Q_{e3} + P_{21}\, Q_{e2} \end{aligned} \quad (1.12)$$

$$\begin{aligned} Q_{u1} &= Q_{21} + Q_{31} + Q_{41} = P_{21}\, Q_{e2} + P_{31}\, Q_{e3} + P_{41}\, Q_{e4} \\ Q_{u2} &= Q_{12} + Q_{32} + Q_{42} = P_{12}\, Q_{e1} + P_{32}\, Q_{e3} + P_{42}\, Q_{e4} \\ Q_{u3} &= Q_{13} + Q_{23} + Q_{43} = P_{13}\, Q_{e1} + P_{23}\, Q_{e2} + P_{43}\, Q_{e4} \\ Q_{u4} &= Q_{14} + Q_{24} + Q_{34} = P_{14}\, Q_{e1} + P_{24}\, Q_{e2} + P_{34}\, Q_{e3} \end{aligned} \quad (1.13)$$

Therefore, it is easy to determine, starting with traffic demand, the circulatory roadway volumes and those traveling towards the exits for a roundabout with a number of legs $i = 1, 2, \ldots$, any n.

Values of the circulating flows Q_{ci} in front of each entry i, $i \in [1, \ldots, n]$ (numbered in an anti-clockwise manner) are

$$\begin{aligned} Q_{c,1} &= Q_{e,n}(P_{n,2} + \ldots + P_{n,n-1}) + Q_{e,n-1}(P_{n-1,2} + \ldots + P_{n-1,n-2}) + \ldots + Q_{e,3}P_{3,2} \\ &\ldots\ldots \\ Q_{c,n} &= Q_{e,n-1}(P_{n-1,1} + \ldots + P_{n-1,n-2}) + Q_{e,n-2}(P_{n-2,2} + \ldots + P_{n-2,n-3}) + \ldots + Q_{e,2}P_{2,1} \end{aligned} \quad (1.14)$$

$$\begin{aligned} Q_{u,1} &= Q_{e,2}P_{2,1} + \ldots + Q_{e,n}P_{n,1} \\ &\ldots\ldots \\ Q_{u,n} &= Q_{e,1}P_{1,n} + \ldots + Q_{e,n-1}P_{n-1,n} \end{aligned} \quad (1.15)$$

1.2 Capacity and Capacity Indices

Capacity C of an entry is defined as the smallest value of the leg flow that causes the permanent presence of vehicles queuing up to enter.

To calculate C, the roundabout is considered as a series, along the development of the junction, of T intersections, with yielding to the circulating flows with only one interaction i.e., the mutual contribution to the formation of circulating flows that affect each entry as conflicting flows.

For a roundabout entry, capacity C may be expressed in the most general way as

$$C = C(\tilde{G}; Q_d; \tilde{\tau}; \tilde{S}) \quad (1.16)$$

where

$\tilde{G}$ is a set of variables representing the geometrical layout of the roundabout (e.g., entry width and central island radius) or its configuration (e.g., number of circulatory roadway lanes and number of entry lanes);

Q_d is the traffic flow that disturbs the traffic entering the roundabout; Q_d is the function of the entry flows (1.1) in general, by means of the circulating flows Q_c (1.8) and the exiting flows Q_u (1.9) (See Fig. 1.1);

$\tilde{\tau}$ is a set of psycho-technical times, relative to user behavior i.e., generally, critical gap T_c and follow-up time T_f[5];

$\tilde{S}$ is a set of numerical constants that result from the calibration process of the capacity formula.

A capacity formula is generally obtained in two ways [4].

a) It may be obtained by the calibration with empirical data of queuing theory models, i.e. models based on the gap-acceptance theory.
b) It may be obtained with empirical regression techniques applied to sampling traffic data without using the queuing theory.

However, specifying Eq. (1.16), today's available[6] capacity formulas may be classified into the following three types:

a) the roundabout is characterized only by its configuration, represented by the number of circle lanes and leg lanes;
b) the geometric design is taken into account at a reasonable level of detail;
c) user behavior with critical gap T_c and follow-up time T_f is taken into account, along with geometric aspects.

Examples of these three types of capacity formulas are dealt with in Chap. 2.

According to some capacity formulas, the disturbing flows Q_d are equal to circulating flows Q_c; according to other formulas, they are equal to appropriate linear combinations of the flows Q_c and Q_u.

Starting with capacity C_i of an entry "i", we will now define further capacity indices that are frequently used to characterize roundabout operational conditions:

– Reserve Capacity $(RC)_i$, which is equal to the difference between capacity C_i and traffic demand Q_{ei} at an entry

[5] T_f is also called move-up time. For the definition of T_c and T_f, see, for example [3].

[6] In the international literature, as far as we know, 32 formulations of capacity have been found, eight of which are German (1992–2007), four are Swiss (1989–2006), two are American (1997–2000), one is English (1980); three are French (1988–1997), three are Dutch (1992–1999); one is Polish (1996), one is Swedish (1996), one is Norwegian (1985), one is Finnish (2004), one is Danish (1999), one is Israeli (1997), two are Australian (1989–1998), two are Austrian (1997–2001), and one is Portuguese (1996).

$$(RC)_i = C_i - Q_{ei} \tag{1.17}$$

– Percentage Reserve Capacity $(RC\%)_i$, given by

$$(RC\%)_i = \frac{(RC)_i}{C_i} \cdot 100 = \frac{C_i - Q_{ei}}{C_i} \cdot 100 \tag{1.18}$$

– Percentage Capacity Rate $(CR\%)_i$, equal to 100 times the traffic intensity $\rho_i = Q_{ei}/C_i$

$$(CR\%)_i = \frac{Q_{ei}}{C_i} \cdot 100 \tag{1.19}$$

We will now give further definitions of capacity related to the entire roundabout:

1) Simple Capacity (SC): with reference to a given traffic demand at the intersection, we define simple capacity SC as the first capacity value that is recorded at an entry for a uniform increase in the flows that make up the demand. In other words, when we calculate simple capacity, we also determine for a given vector (1.1) and matrix (1.2) the value of the maximum entering flow on each leg when one of the entry, in the uniform evolution of the system, becomes saturated.
2) Total Capacity (TC): we define total capacity TC with respect to a given percentage distribution of entering traffic (See matrix (1.2)) the sum of the flows from each entry (Q_{ei}) that are simultaneously equal to capacity ($Q_{ei} = C_i$).

Therefore, Total Capacity indicates, for a given distribution of demand at the intersection (i.e., for a given determination of (1.2)), a concise measurement of the roundabout limited ability to serve traffic when each leg is saturated.

The procedures to determine SC and TC will be illustrated in Chap. 2.

Finally, capacity indices (absolute and in percentage) for the entire roundabout may be defined as the weighted means on the flows of the determinations obtained, respectively, with Eqs. (1.17), (1.18), and (1.19) for each roundabout entry.

1.3 Delays and Queue Lengths

Delays at intersections along the route contribute to lost time during the travel.

Figure 1.3 shows the components that make up the delay due to the presence of a roundabout along the route.

The simplified time-distance diagram shown in Fig. 1.4 may replace the time-distance diagram shown in Fig. 1.3 in order to evaluate delays.

In Fig. 1.4, the horizontal line w_s indicates, using the term of the queuing theory [5], the time spent in the system or the delay of the vehicle in the sense of traffic engineering.

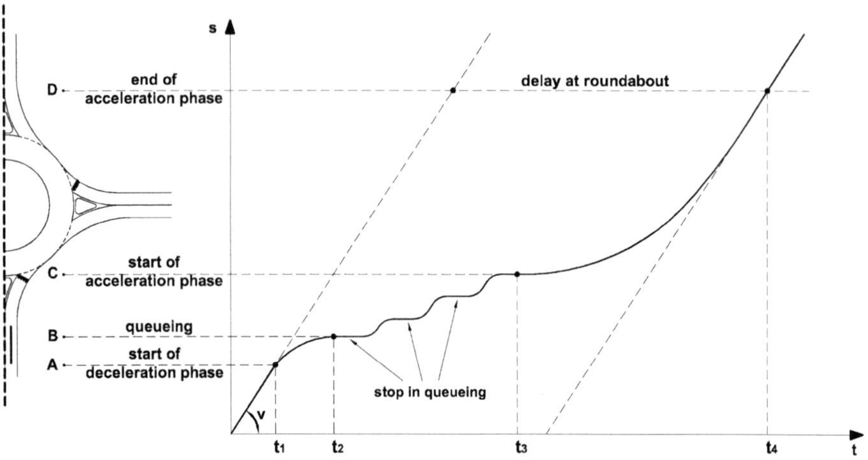

Fig. 1.3 Delays due to the presence of a roundabout along the route

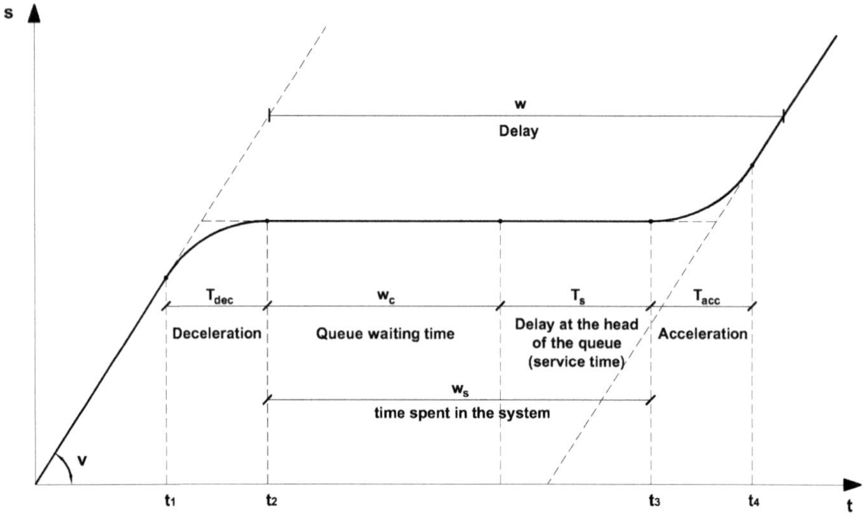

Fig. 1.4 Simplified space versus time diagram

It results from the addition of two rates:

- queue waiting time w_c, i.e., the time that a single vehicle spends from the moment it joins the queue till it gets to the head of the queue (i.e., to the yielding line);
- service time T_S, which is the time between the moment the vehicle reaches the head of the queue (i.e., the yielding line) and the moment it enters the circulatory roadway to perform the desired maneuver.

Finally,

$$w_s = w_c + T_s \tag{1.20}$$

Now, with the model generally used to obtain the delay w (total delay) (See Fig. 1.4), the times T_{dec} and T_{acc} must be considered together with w_s. They indicate the deceleration phase when approaching the intersection and the acceleration phase when crossing and exiting the intersection, respectively.

In the roundabout of Fig. 1.4, $T_{dec} = t_2 - t_1$, and $T_{acc} = t_4 - t_3$.

Thus, the delay at the intersection w, is function, among other things, of

$$w = f(w_s;T_{dec};T_{acc};(.);(.)...) \tag{1.21}$$

In conclusion, total delay is the difference between the traveling time spent in presence of the intersection and the traveling time that would be spent in absence of the intersection.

The difference w_g

$$w_g = w - w_s \tag{1.22}$$

is called geometric delay.

The determination of w is used, for example: (a) to evaluate costs related to traffic assignment to road networks; (b) to make the comparative analysis used to change the layout of an intersection; (c) to compare roundabouts of different geometry.

To evaluate the quality of circulation at given intersections, it is, instead, sufficient to calculate only the time spent in the system w_s provided by Eq. (1.20).

We will henceforth analyze only w_s.[7]

However, to calculate geometric delay, see, for example [6].

L_s at an entry indicates the number of waiting vehicles in the system, i.e., including the vehicle in service[8] (See Fig. 1.5).

The length of the queue L_c is made up, instead, of the number of vehicles waiting behind the vehicle in service, e.g., in the case of a roundabout with single-lane entries, $L_c = L_s - 1$.

L_s is the number of users in the system.
w_s, w_c, L_s, and L_c are random variables.

[7]In technical practice, but sometimes also in scientific literature, "delay" does not mean only w given by Eq. (1.21), but also w_s expressed by Eq. (1.20) that, as we have already pointed out, is part of w and is relative only to the time spent queuing up at the beginning of the maneuver to enter the circle.

[8]The vehicle in service is the first vehicle in the queue, i.e., the nearest to the yielding line, waiting to enter the circle.

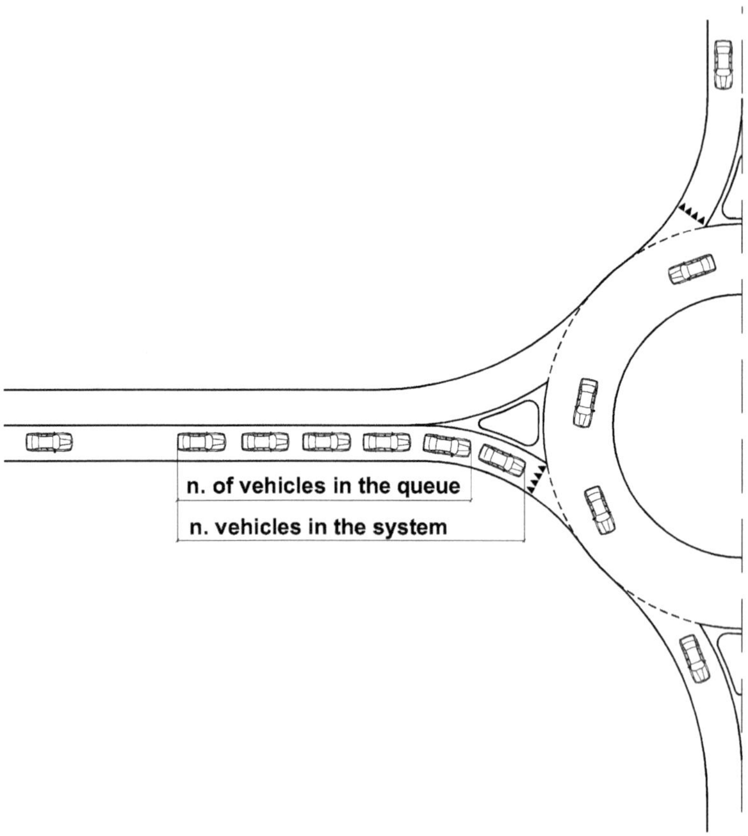

Fig. 1.5 Number of vehicles in the system and number of vehicles in the queue

To determine the effectiveness measures of roundabouts, we need the mean values $E[w_S]$ of w_S and $E[w_C]$ of w_C and, for L_S and L_C, in addition to the means $E[L_S]$ and $E[L_C]$, we need suitable percentiles $L_{S,p}$ and $L_{C,p}$. The order p of the percentiles may be established in relation to the particular problem under examination.

As we have already pointed out in Sect. 1.1, the determinations of $E[w_S]$, $E[w_C]$, $E[L_S]$, $E[L_C]$, $L_{S,p}$, $L_{C,p}$ require the characterization of the state of the system in terms of the presence or absence of steady-state conditions.

In a steady-state condition, the mean (average) $E[\cdot]$ is a mathematical expectation of a random variable.

To define, instead, the mean in the absence of steady-state conditions, it is necessary to take into account that, under these conditions, the queue mean length varies according to the change of the system conditions (traffic demand variation and/or entry states).

In particular, if the absence of a steady-state condition is connected to saturation or oversaturation at the entry of concern for a period T, at the beginning of T, queues shorter than those at the end of the same period will be recorded.

For the time spent in the system and waiting time in the queue, it follows that the same behavior, i.e. a shorter duration at the beginning of T compared to that recorded at its end, may be recorded.

Thus, for example, the average waiting time in the queue $E[w_c]$ related to T is defined as

$$E[w_c] = \frac{\overline{w}_c(t_0) + \overline{w}_c(t)}{2} \tag{1.23}$$

where $\overline{w}_c(t_0)$ and $\overline{w}_c(t)$ are the waiting times in the queue of two vehicles joining the queue at the beginning t_0 and at the end $t = t_0 + T$ of the observation period T.

Equation (1.23) may be also considered as an estimate of the time that an user may spend, on average, in the queue when he or she arrives at half of the period of entry saturation or oversaturation.

In addition, it is important for practical applications to evaluate the percentile $L_{s,p}$ of the number of users in the system at the end of T, for the situations connected to saturated or oversaturated entries.

The estimation of a suitable $L_{s,p}$ in an entry to a roundabout is an essential task for the following reasons:

a) it is an indicator of traffic quality.
b) it is useful to design and validate the geometric layout of the intersection.
c) it helps to prevent traffic jams into upstream intersections.

Chapter 3 will be dedicated to the criteria to calculate the waiting phenomena parameters at roundabout entries.

References

1. Hsu, H. W., *Probability, Random Variables and Random Processes*, MacGraw-Hill, New York, 1997.
2. Roess, P. R., McShane, R. W., Prassas, S. E., *Traffic Engineering* – 2nd Ed., Prentice Hall, Upper Saddle River, New Jersey, 1998.
3. Louah, G., *Priority Intersections: Modelling*, in “Concise Encyclopedia of Traffic and Transportation System” (Papageorgiu, M., Editor), Pergamon Press, Oxford, 1991.
4. Brilon, W. (Editor), *Intersections without Traffic Signals* – Vol. I and II, Springer-Verlag, Heidelberg, 1998 and 1991.
5. Kleinlock, L., *Queuing Systems* – Vol. I, John Wiley and Sons, New York, 1975.
6. Teakratok, T., *Modern Roundabout for Oregon*, Oregon Department of Research Unit, 1998.

Chapter 2
Capacity Evaluation

In the previous Sect. 1.2, the definitions of capacity C of an entry, simple capacity SC, and whole or total capacity TC were discussed.

The section also contains further definitions regarding reserve capacity indices (absolute RC, percentage RC%, and percentage Capacity Rate CR%).

As we will see in the following Sect. 2.1, the determination of Capacity C_i at the entries "i" is easy if the entries are all at undersaturation conditions $\left((RC)_i = C_i - Q_{ei} > 0, \text{i.e.}, \rho_i = Q_{ei}/C_i < 1\,\forall i\right)$.

On the other hand, if one or more of the entries "i" are not undersaturated, the determination of their capacities C_i is not straightforward. The determinations of simple capacity SC and total capacity TC are not straightforward.

In all these cases, it is necessary to use specific, iterative calculation methods, which will be dealt with in the following Sects. 2.5 and 2.6.

2.1 Capacity Calculation at Steady-State Conditions

As already described, by specifying Eq. (1.16) in Chap. 1, the capacity formulas that are currently available may be classified into three types:

a) the roundabout is characterized only by its configuration, represented by the number of circle lanes and leg lanes;
b) the roundabout geometry is taken into account in somewhat detailed way;
c) we take into account, together with geometric aspects, the users' behaviors thanks to psycho-technical times T_c, critical gap, and follow-up time T_f.

To provide some examples of the type of relationships of the above-mentioned classifications (a; b; c), six capacity formulas are now presented and implemented.

Two formulas of type (a) that were developed by Brilon et al. (Germany) and by Bovy et al. (Switzerland) are presented in which the roundabout configuration is therefore represented by the number of lanes at the entries and in the circle. In addition, according to Brilon's formula, the disturbing flow Q_d is represented only

R. Mauro, *Calculation of Roundabouts*, DOI 10.1007/978-3-642-04551-6_2,

by the circulating flow Q_c in front of the entries, whereas, according to Bovy's formula, Q_d is a linear combination of Q_c and exiting traffic Q_u.[1]

The third formula presented, developed by TRRL (United Kingdom), belongs to set (b) and requires, with respect to the two previous cases, a more detailed characterization of the roundabout geometry and uses the size of the main planimetric elements of the roundabout as input; in this formula, the disturbing flow is represented only by the circulating flow Q_c.

The fourth formula presented, the GIRABASE procedure (France), belongs to set (c). In fact, it is necessary to determine some geometric values of the roundabout and to use a pre-fixed follow-up time value T_f to implement the formula. In this case, the disturbing flow is represented by a linear combination of the circulating flow with the exiting flow.

The fifth formula, developed by Brilon and Wu (Germany), belongs to set (c), and it considers capacity as a function of the roundabout configuration, rendered in terms of number of lanes at the entries and in the circle and as function of the users' psycho-technical attitudes, expressed by determining the critical gap T_c and follow-up time T_f[2] values.

In this case, Q_d is set equal to Q_c.

The sixth formula also belongs to set (c). It is included in the Highway Capacity Manual HCM 2002 [1].

The capacity formulas available in the literature generally express entering flows in passenger car unities (pcu/h), and, using the same measure, entry capacities are obtained.

To express flows Q_e in pcu/h, vehicles other than passenger cars are generally treated as follows: 1 truck, bus = 1.5 pcu; 1 truck + trailer = 2.0 pcu; 1 motorcycle = 0.5 pcu; 1 bicycle = 0.5 pcu.

When formulas make use of the number of lanes, one should reason as follows. According to regulations that are largely accepted, circle lanes must not have road markings; for this reason, the phrase "number of circle lanes" is meant to be the number of circulating vehicles rows that can be accommodated on the circulatory roadway.

In all the following examples, we assume that all the intersections under examination are at steady-state conditions.

We wish to recall (see Sect. 1.1) that a steady-state condition, as we mean it here, is achieved if entries are undersaturated and traffic demand at each leg remains constant for a time period T of a suitable size. In other words, T must be long enough to allow the operating conditions of the intersection to become steady with constant mean values of state variables. In addition, the punctual values of state variables must be little dispersed around the mean values.

---

[1] For Q_e, Q_u, and Q_c, see Fig. 1.2 in Chap. 1.

[2] To evaluate T_c and T_f, see, for example, [3].

2.1.1 Brilon-Bondzio Formula (Germany)

The capacity of an entry is represented by the simple linear relationship [2, 3]

$$C = A - B \cdot Q_c \ (\text{pcu/h}) \tag{2.1}$$

where A and B are obtained from Table 2.1, depending on the numbers of entry and circle lanes.

Equation (2.1) is valid for roundabouts with external diameters D_{ext} that range from 28 to 100 m.

In Eq. (2.1), the disturbing traffic Q_d coincides with the circulating flow Q_c in front of the entry for which C is determined.

Figure 2.1 shows the representation of Eq. (2.1) for all the geometric configurations for which it is valid.

As an example, consider a four-legged roundabout with a double-lane circle and double-lane entries.

Table 2.1 Parameters values for Brilon-Bondzio capacity formula

Circle lane number	Entry lane number	A	B	Sample size
3	2	1409	0.42	295
2	2	1380	0.50	4574
2–3	1	1250	0.53	879
1	1	1218	0.74	1504

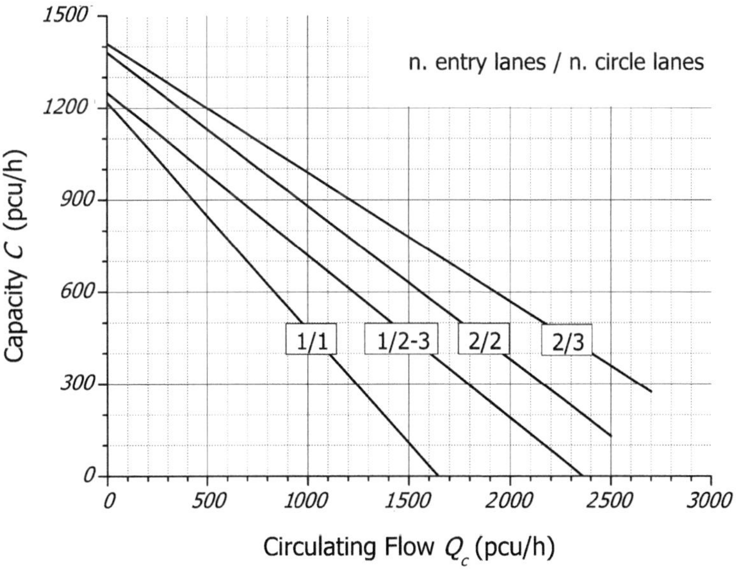

Fig. 2.1 Capacity C versus circulating flow Q_c according to Brilon-Bondzio capacity formula

The entering flow vector (pcu/h) is $Q_e = [322, 252, 329, 408]$.

The origin/destination matrix relative to the time considered is the following (the traffic volumes in matrix $M_{O/D}$ are expressed in pcu/h):

$$M_{O/D} \equiv \begin{bmatrix} 0 & 82 & 116 & 124 \\ 74 & 0 & 92 & 86 \\ 106 & 96 & 0 & 127 \\ 128 & 141 & 139 & 0 \end{bmatrix}$$

From the entering flows vector Q_e, with Eqs. (1.12) and (1.13) in Chap. 1, we obtain the exiting flows (Q_u) and circulating flows in front of each entry (Q_c) for each leg, i.e., i = 1, 2, 3, 4, of the roundabout:

$$\begin{aligned} Q_{u1} &= 308\,\text{pcu/h} & \qquad Q_{c1} &= 376\,\text{pcu/h} \\ Q_{u2} &= 319\,\text{pcu/h} & Q_{c2} &= 379\,\text{pcu/h} \\ Q_{u3} &= 347\,\text{pcu/h} & Q_{c3} &= 284\,\text{pcu/h} \\ Q_{u4} &= 337\,\text{pcu/h} & Q_{c4} &= 276\,\text{pcu/h} \end{aligned}$$

On the basis of Eq. (2.1), with A = 1380 and B = 0.50 (See Table 2.1), the capacities for each entry are:

$$\begin{aligned} C_1 &= 1380 - 0.5 \cdot Q_{c1} = 1192\,\text{pcu/h} \\ C_2 &= 1380 - 0.5 \cdot Q_{c2} = 1190\,\text{pcu/h} \\ C_3 &= 1380 - 0.5 \cdot Q_{c3} = 1238\,\text{pcu/h} \\ C_4 &= 1380 - 0.5 \cdot Q_{c4} = 1242\,\text{pcu/h} \end{aligned}$$

Then, we evaluate the capacity indices for each entry (See Sect. 1.2).

The reserve capacities are:

$$\begin{aligned} (RC)_1 &= C_1 - Q_{e1} = 1192 - 322 = 870\,\text{pcu/h} \\ (RC)_2 &= C_2 - Q_{e2} = 1190 - 252 = 938\,\text{pcu/h} \\ (RC)_3 &= C_3 - Q_{e3} = 1238 - 329 = 909\,\text{pcu/h} \\ (RC)_4 &= C_4 - Q_{e4} = 1242 - 408 = 834\,\text{pcu/h} \end{aligned}$$

Percentage Capacity Rates (CR%) for each entry are

$$\begin{aligned} (CR\%)_1 &= (Q_{e1}/C_1) \cdot 100 = (322/1192)\,100 = 27.0\% \\ (CR\%)_2 &= (Q_{e2}/C_2) \cdot 100 = (252/1190)\,100 = 21.2\% \\ (CR\%)_3 &= (Q_{e3}/C_3) \cdot 100 = (329/1238)\,100 = 26.6\% \\ (CR\%)_4 &= (Q_{e4}/C_4) \cdot 100 = (408/1242)\,100 = 32.9\% \end{aligned}$$

For the roundabout as a whole, the following mean values are obtained

$$\overline{RC}=\frac{\sum_{i=1}^{n}(RC)_i\cdot Q_{ei}}{\sum_{i=1}^{n}Q_{ei}}=\frac{870\cdot 322+938\cdot 252+909\cdot 329+834\cdot 408}{322+252+329+408}\cong 881\ \text{pcu/h}$$

$$\overline{(CR\%)}=\frac{\sum_{i=1}^{n}(CR\%)_i\cdot Q_{ei}}{\sum_{i=1}^{n}Q_{ei}}=$$

$$=\frac{27.0\%\cdot 322+21.2\%\cdot 252+26.6\%\cdot 329+32.9\%\cdot 408}{322+252+329+408}\cong 27.6\%$$

2.1.2 Bovy et al. Formula (Switzerland)

This formula is recommended for roundabouts (to be used in urban and suburban environments) with a non mountable central island, of small dimensions (maximum internal diameter D_{int} = 18 – 20 m) [3]. The circle external diameter D_{ext} varies generally from 24 to 34 m, and there are flared entries, i.e., there are more lanes next to the stop line to make the choice of the desired direction easier.

An entry capacity is determined with the relationship:

$$C=\frac{1}{\gamma}\cdot\left(1500-\frac{8}{9}\cdot Q_d\right)\ (\text{pcu/h}) \tag{2.2}$$

where γ is a parameter that allows taking into account the number of entry lanes, and its value is:

- $\gamma = 1$ for one lane;
- $\gamma = 0.6$–0.7 m for two lanes (according to smaller or larger entry dimensions), and it is generally set at 0.667;
- $\gamma = 0.5$ for three lanes.

Q_d is the disturbing traffic determined as:

$$Q_d=\alpha\cdot Q_u+\beta\cdot Q_c\ (\text{pcu/h}) \tag{2.3}$$

where (See Fig. 2.2):

Q_u = exiting traffic;
Q_c = circulating traffic in front of the exit being considered.

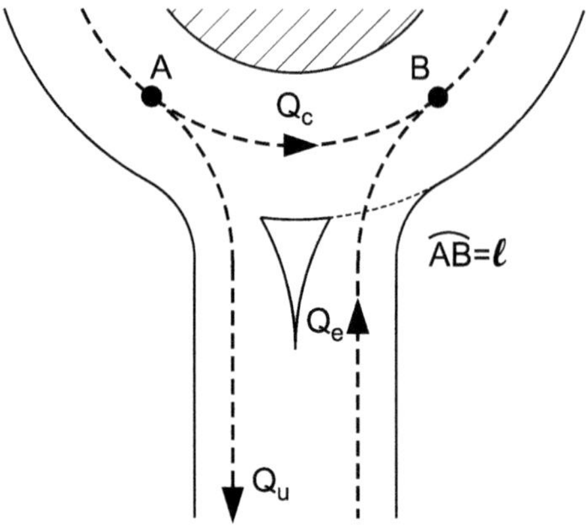

Fig. 2.2 Distance ℓ between the exiting conflicting point (**A**) and entering point (**B**)

Coefficients α and β are related to the geometry of the roundabout and take into account the distance ℓ between the exiting conflicting points (A) and entering points (B) that are conventionally identifiable on the circle (Fig. 2.2) and the number of circular lanes, respectively.

Due to the ability to simulate the roundabout, it was possible to establish that, in accordance with the intuitively obvious inverted proportionality relationship between ℓ and the percentage disturb due to the vehicles exiting the intersection, α decreases with ℓ until, for $\ell > 28$ m, the exiting vehicles do not disturb the entering vehicles ($\alpha = 0$).

Figure 2.3 shows three behaviors of the value α as a function of the distance ℓ. The line "a" is relative to a circle flow speed of 20–25 km/h; lines "b" and "c" border the band above and below "a" when V > 20–25 km/h (greater disturb) and when V < 20–25 km/h (smaller disturb), respectively.

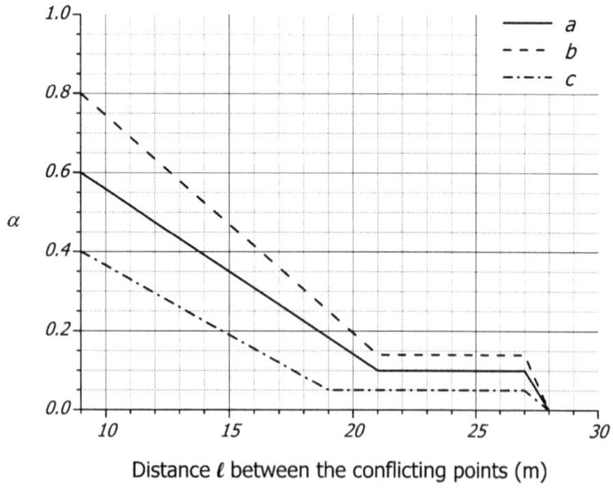

Fig. 2.3 Values of parameter α versus the distance ℓ between the exiting and entering conflicting points shown in Fig. 2.2

For β, which takes into account the reduction effect that the presence of more than one circle lane has on the conflicting flow caused by the circulating traffic, the following values are provided: $\beta = 0.9$–1.0 for one lane; $\beta = 0.6$–0.8 for two lanes; and $\beta = 0.5$–0.6 for three lanes.

To calculate the number of equivalent passenger car units (pcu), the following values are suggested:

one bike or motorbike in the circle − 0.8 pcu
one entering bike or motorbike = 0.2 pcu
one heavy vehicle or bus = 2.0 pcu

The Swiss Standards on roundabouts use the formula proposed by Bovy et al. and use the following two capacity indices as the roundabout efficiency indicators. The calculation of these indices and their meanings are straightforward, and they must always be determined together. The calculations are performed as shown below:

– Capacity Rate used at entries (CRU_e):

$$CRU_e = (\gamma \cdot Q_e/C_e) \cdot 100\ (\%) \tag{2.4}$$

– Capacity Rate used at the conflicting point (CRU_c):

$$CRU_c = [(\gamma \cdot Q_e + 8/9 \cdot Q_d)/1500] \cdot 100\ (\%) \tag{2.5}$$

Now, we present an example of implementation of the procedure for the roundabout examined in the previous Sect. 2.1.1 (a four-legged roundabout with a double-lane circle and double-lane entries). We assume that the distance ℓ between the conflicting points is equal to 20 m in front of each leg.

The traffic data are the same as the traffic data used in the example in Sect. 2.1.1.

The circle flow speed is assumed to be 20–25 km/h. The value of the parameter α is determined from the diagram shown in Fig. 2.3, along the line "a", as a function of the distance $\ell = 20$ m for all the legs. The result is $\alpha = 0.14$.

Since the circle has two lanes, β is equal to 0.7.

Thus, disturbing flows in front of the legs can be determined in the following way:

$$\begin{aligned}
Q_{d1} &= \alpha \cdot Q_{u1} + \beta \cdot Q_{c1} = 0.14 \cdot 308 + 0.7 \cdot 376 \cong 307\,\text{pcu/h}\\
Q_{d2} &= \alpha \cdot Q_{u2} + \beta \cdot Q_{c2} = 0.14 \cdot 319 + 0.7 \cdot 379 \cong 311\,\text{pcu/h}\\
Q_{d3} &= \alpha \cdot Q_{u3} + \beta \cdot Q_{c3} = 0.14 \cdot 347 + 0.7 \cdot 284 \cong 249\,\text{pcu/h}\\
Q_{d4} &= \alpha \cdot Q_{u4} + \beta \cdot Q_{c4} = 0.14 \cdot 337 + 0.7 \cdot 276 \cong 242\,\text{pcu/h}
\end{aligned}$$

Therefore, with Eq. (2.2), we can determine the capacity values at entries, using $\gamma = 0.667$ for all (because of double-lane entries), as shown below:

$$C_1 = 1/\gamma \cdot (1500 - 8/9 \cdot Q_{d1}) = 1/0.667 \cdot (1500 - 8/9 \cdot 307) \cong 1840\,\text{pcu/h}$$
$$C_2 = 1/\gamma \cdot (1500 - 8/9 \cdot Q_{d2}) = 1/0.667 \cdot (1500 - 8/9 \cdot 311) \cong 1834\,\text{pcu/h}$$
$$C_3 = 1/\gamma \cdot (1500 - 8/9 \cdot Q_{d3}) = 1/0.667 \cdot (1500 - 8/9 \cdot 249) \cong 1917\,\text{pcu/h}$$
$$C_4 = 1/\gamma \cdot (1500 - 8/9 \cdot Q_{d4}) = 1/0.667 \cdot (1500 - 8/9 \cdot 242) \cong 1926\,\text{pcu/h}$$

Finally, the values CRU_e and CRU_c are:

$$CRU_{e1} = (\gamma \cdot Q_{e1}/C_{e1}) \cdot 100 = (0.667 \cdot 322/1840)\,100 = 11.7\%$$
$$CRU_{e2} = (\gamma \cdot Q_{e2}/C_{e2}) \cdot 100 = (0.667 \cdot 252/1834)\,100 = 9.2\%$$
$$CRU_{e3} = (\gamma \cdot Q_{e3}/C_{e3}) \cdot 100 = (0.667 \cdot 329/1914)\,100 = 11.4\%$$
$$CRU_{e4} = (\gamma \cdot Q_{e4}/C_{e4}) \cdot 100 = (0.667 \cdot 408/1926)\,100 = 14.1\%$$
$$CRU_{c1} = (\gamma . Q_{e1} + 8/9 \cdot Q_{d1})/1500 \cdot 100 = (0.667 \cdot 322 + 8/9 \cdot 307)/1500 \cdot 100 = 32.5\%$$
$$CRU_{c2} = (\gamma . Q_{e2} + 8/9 \cdot Q_{d2})/1500 \cdot 100 = (0.667 \cdot 252 + 8/9 \cdot 311)/1500 \cdot 100 = 29.6\%$$
$$CRU_{c3} = (\gamma . Q_{e3} + 8/9 \cdot Q_{d3})/1500 \cdot 100 = (0.667 \cdot 329 + 8/9 \cdot 249)/1500 \cdot 100 = 29.4\%$$
$$CRU_{c4} = (\gamma . Q_{e4} + 8/9 \cdot Q_{d4})/1500 \cdot 100 = (0.667 \cdot 408 + 8/9 \cdot 242)/1500 \cdot 100 = 32.5\%$$

2.1.3 TRRL Formula (United Kingdom)

With the TRRL formula, capacity C of a generic entry is determined as a function of the leg and circle geometric parameters and of the circulating flow in the circle (Q_c) in front of the entry [4].

The relationship was developed by Kimber, and it is based on experimental observations of a large number of operating roundabouts in England. It has the following linear form:

$$C = k \cdot (F - f_c Q_c)\,(\text{pcu/h}) \tag{2.6}$$

where:

$$F = 303 \cdot x_2$$
$$f_c = 0.210 \cdot t_D \cdot (1 + 0.2 \cdot x_2)$$
$$k = 1 - 0.00347 \cdot (\Phi - 30) - 0.978 \cdot (1/r - 0.05)$$

$$t_D = 1 + \frac{1}{2 \cdot [1 + \exp((D - 60)/10)]}$$

$$x_2 = v + \frac{(e - v)}{(1 + 2 \cdot S)}$$

$$S = 1.6\,(e - v)/\ell' = (e - v)/\ell$$

Table 2.2 Geometric parameters used by the TRRL formula

Parameter	Description	Range values
e	Entry width	3.6–16.5 m
v	Lane width	1.9–12.5 m
e′	Previous entry width	3.6–15.0 m
v′	Previous lane width	2.9–12.5 m
u	Circle width	4.9–22.7 m
ℓ, ℓ'	Flare mean length	1–∞ m
S	Sharpness of the flare	0–2–9
r	Entry bend radius	3.4–∞ m
Φ	Entry angle	0–77°
$D = D_{ext}$	Inscribed circle diameter	13.5–171.6 m
W	Exchange section width	7.0–26.0 m
L	Exchange section length	9.0–86.0 m

Table 2.2 shows the geometric parameters, the respective symbols used in the procedure, and their range [4].

The main indications contained in [4] for the determination of such parameters are now presented. However, this determination can sometimes be rather difficult because of the particular geometric configurations of the roundabout.

To illustrate the geometric elements that are used in the formula, it is useful to observe Figs. 2.4, 2.5, and 2.6. They are taken from the original work [4], and they show the left-side driving in the United Kingdom requiring that travel in the circle be clockwise. To apply the procedure to counterclockwise roundabouts that are used for right-side traffic, homologous symmetrical elements must be used.

The width of the entry (e) is determined along the perpendicular line traced from point A to the external edge (See Fig. 2.4).

The width of the entry lane (v) must be determined upstream of the leg widening next to the entry along the perpendicular line traced from the axis of the roadway to the external edge.

The width of the circulatory roadway (u) represents the distance between the splitter island at legs (point A) and the central island.

The entry radius (r) is the smallest bend radius of the external edge next to the entry.

The width of the weaving section (W) is the shortest distance between the central island and the external edge in the stretch between an entry and the following exit.

The weaving section (L) is defined as the shortest distance between the splitter islands at the legs of two successive entries.

The mean length of the flare can be determined using either of the two parameters ℓ or ℓ'. Figure 2.5 shows the geometric constructions for their determination. In both cases, by tracing a line parallel to the curve HA (at a distance v from it), we can determine the curve GD, which intersects the segment AB (which represents the entry width) at point D; the length ℓ corresponds to the segment CF, determined along the perpendicular line that passes through C (mean point of segment BD) of segment AB (with F as intersection point between the above-mentioned perpendicular and the curve GD); length ℓ' corresponds to segment CF′ along a curve parallel to

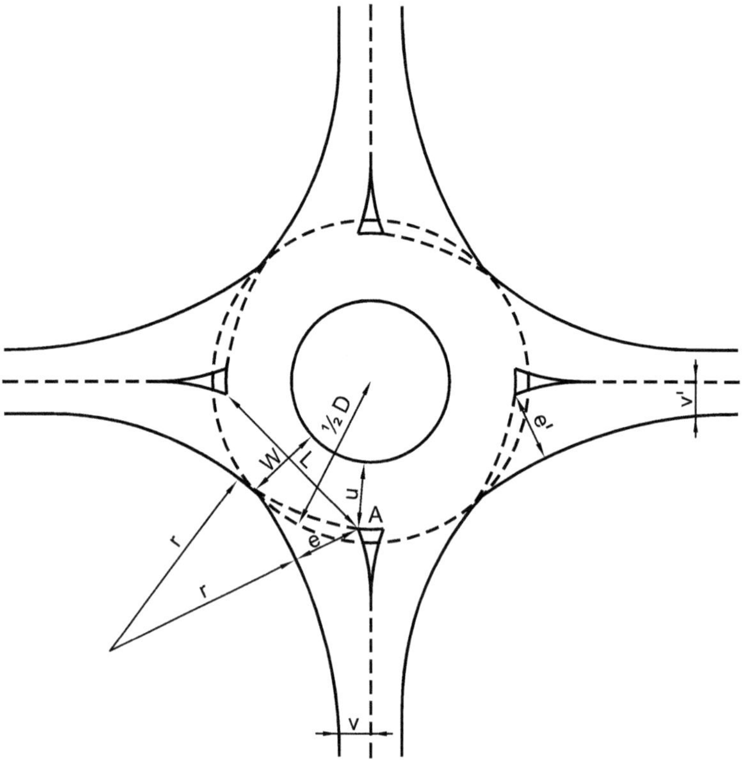

Fig. 2.4 Geometric elements used in the TRRL formula [3]

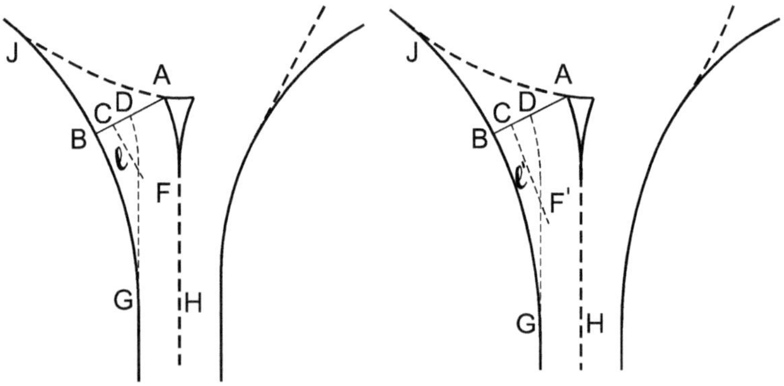

Fig. 2.5 Geometric construction for the determination of ℓ and ℓ' [3]

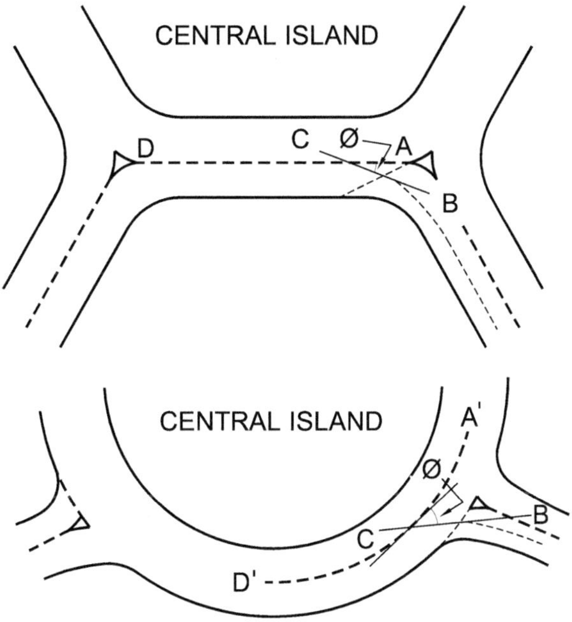

Fig. 2.6 Geometric construction for the determination of the entry angle Φ [4]

the external edge BG and passing through C (with F′ as intersection point between the above-mentioned curve and the curve GD).

Between ℓ and ℓ', the approximate relationship $\ell' = 1.6\,\ell$ is valid within the allowed variability for the practical use of geometric parameters.

The entry angle (Φ), which represents the conflicting angle between the entering flows and the circulating flows, must be determined according to the straightforward indications shown in Fig. 2.6.

Now, we present an example of the application of the procedure.

Consider a four-legged roundabout with the following features:

– inscribed circle diameter: D = 45 m;
– entry width: e = 4.8 m;
– lane width: v = 3.5 m;
– flare mean length $\ell' = 25$ m;
– entry radius r = 40 m;
– entry angle: $\Phi = 60°$.

Traffic data (expressed in pcu/h) are summarized in the matrix O/D:

$$M_{O/D} \equiv \begin{bmatrix} 0 & 150 & 300 & 200 \\ 200 & 0 & 150 & 350 \\ 350 & 150 & 0 & 150 \\ 300 & 250 & 200 & 0 \end{bmatrix}$$

From matrix O/D, the circulating flows in front of each entry can be obtained (Eq. (2.12)):

$$\begin{aligned} Q_{c1} &= 600\,\text{pcu/h} \\ Q_{c2} &= 700\,\text{pcu/h} \\ Q_{c3} &= 750\,\text{pcu/h} \\ Q_{c4} &= 700\,\text{pcu/h} \end{aligned}$$

Then, we can determine the calculation parameters of each leg using Eq. (2.6) since all of the geometric parameters relative to each entry are the same:

$$\begin{aligned} & S = 1.6 \cdot (e - v)/\ell' = 1.6 \cdot (4.8 - 3.5)/25 = 0.083 \\ & x_2 = v + (e - v)/(1 + 2 \cdot S) = 3.5 + (4.8 - 3.5)/(1 + 2 \cdot 0.083) = 4.615 \\ & F = 303 \cdot x_2 = 303 \cdot 4.615 = 1398 \\ & t_D = 1 + 0.5/[1 + \exp((D - 60)/10)] \\ & \quad = 1 + 0.5/[1 + \exp((45 - 60)/10)] = 1.409 \\ & f_c = 0.210 \cdot t_D \cdot (1 + 0.2 \cdot x_2) = 0.210 \cdot 1.409 \cdot (1 + 0.2 \cdot 4.615) = 0.569 \\ & k = 1 - 0.00347 \cdot (\Phi - 30) - 0.0978 \cdot (1/r - 0.05) = \\ & \quad = 1 - 0.00347.(60 - 30) - 0.978 \cdot (1/40 - 0.05) = 0.920 \end{aligned}$$

Therefore the capacity formula for all the legs can be written as:

$$C = k \cdot (F - f_c Q_c) = 0.920 \cdot (1398 - 0.569 \cdot Q_c) = 1286 - 0.523 \cdot Q_c$$

The capacity values determined for the four entries are:

$$\begin{aligned} C_1 &= 972\ \text{pcu/h} \\ C_2 &= 919\ \text{pcu/h} \\ C_3 &= 893\ \text{pcu/h} \\ C_4 &= 919\ \text{pcu/h} \end{aligned}$$

Among the further applications of the TRRL capacity formula, we illustrate the determination of a roundabout entry width.

The entering flow of an entry is set to $Q_{ei} = 800$ pcu/h; the circulating flow in front of the entry is set to $Q_{ci} = 1100$ pcu/h.

The geometric data relative to the entry are the following:

- inscribed circle diameter $D = 40$ m;
- entry lane width $v = 7.3$ m;
- flare mean length $\ell' = 20$ m;
- entry bend radius $r = 25$ m; and
- entry angle $\Phi = 30°$.

We want to determine the entry width "e" necessary, for example, to ensure a reserve capacity equal to the entering flow.

Therefore, we have:

$$k = 1 - 0.00347 \cdot (30 - 30) - 0.978 \cdot (1/25 - 0.05) = 1.01$$
$$t_D = 1 + 0.5/[1 + \exp((40 - 60)/10)] = 1.44$$
$$f_c = 0.210 \cdot t_D \cdot (1 + 0.2 \cdot x_2) = 0.210 \cdot 1.44 \cdot (1 + 0.2 \cdot x_2)$$
$$F = 303 \cdot x_2$$

The entry capacity necessary to have a reserve capacity equal to the value of the entering flow is equal to $2 \cdot 800 = 1600$ pcu/h.

Therefore, we have:

$$C = k \cdot (F - f_c Q_c)$$
$$1600 = 1.01 \cdot [303 \cdot x_2 - 0.210 \cdot 1.44 \cdot (1 + 0.2 \cdot x_2) \cdot 1100]$$
$$1600 = 306.03 \cdot x_2 - 335.97 - 67.19 \cdot x_2$$
$$1935.97 = 238.84 \cdot x_2$$
$$x_2 = 8.11$$

On the other hand, we can also determine:

$$S = 1.6(e - v)/\ell' = 1.6(e - 7.3)/20$$
$$x_2 = v + \frac{(e - v)}{(1 + 2 \cdot S)} = 7.3 + \frac{(e - 7.3)}{[1 + 2 \cdot 1.6 \cdot (e - 7.3)/20]}$$

Equating the two expressions for x_2, we obtain the value of the width requested, which is $e = 8.23$ m.

2.1.4 GIRABASE Formula (France)

GIRABASE is the commercial software currently used in France to determine the capacity of a roundabout. It was developed by CETE de l'Ouest of Nantes and was accepted by CERTU and by SETRA[3] [5].

The final version of the formula was written after it was tested by Urbahn in Germany in 1996 and updated by Guichet in 1997.

It was developed by treating traffic data collected by observing the entries of roundabouts (operating at saturation conditions) with statistical regression techniques. In particular, GIRABASE software's empirical regression equations are based on the counting of 63000 vehicles during 507 saturated operation periods of 5–10 min in 45 different roundabouts [6].

The procedures can be used for all types of roundabouts (from small roundabouts to large roundabouts) located in urban or rural areas, with the number of legs ranging from three to eight and with one, two, or three circle lanes and entry lanes [7].

[3] CETE: Centre d'Etudes Techniques de l'Equipement; CERTU: Centre d'Etudes sur les Réseaux, les Transport, l'Urbanisme et les constructions publiques; SETRA: Service d'Etudes Techniques des Routes et Autoroutes.

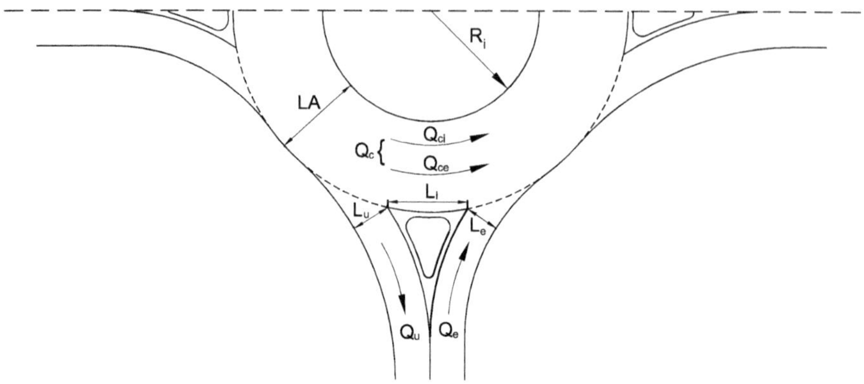

Fig. 2.7 Traffic flows and geometric elements for the GIRABASE formula

Figure 2.7 shows the traffic flows and the geometric elements of the roundabout considered in the procedure; Table 2.3 shows the ranges of the geometric elements for the application of the procedure.

The formula for the determination of entry capacity (pcu/h), based on the exponential regression technique, is the following:

$$C = A \cdot e^{-C_B \cdot Q_d} \tag{2.7}$$

with

$$A = \frac{3600}{T_f} \left(\frac{L_e}{3.5} \right)^{0.8} \tag{2.8}$$

T_f = follow-up time = 2.05 s;
L_e = width of the entry in proximity to the roundabout, determined perpendicularly to the entry direction (m);
C_B = coefficient that is 3.525 for urban areas and 3.625 for rural areas;

Table 2.3 Range of the roundabout geometric elements for the application of the GIRABASE procedure

Parameter	Description	Range values
L_e	Entry width	3–11 m
L_i	Splitter island width	0–70 m
L_u	Exit width	3.5–10.5 m
LA	Circle width	4.5–17.5 m
R_i	Central island radius	3.5–87.5 m

$$Q_d = Q_u \cdot k_a \cdot \left(1 - \frac{Q_u}{Q_c + Q_u}\right) + Q_{ci} \cdot k_{ti} + Q_{ce} \cdot k_{te} \tag{2.9}$$

Q_d = disturbing flow in front of the entry (pcu/h);
Q_u = exiting flow (pcu/h);
$Q_c = Q_{ci} + Q_{ce}$ = circulating flow in front of the entry (pcu/h);
Q_{ci} = traffic rate Q_c on the inner circle lane (pcu/h);
Q_{ce} = traffic rate Q_c on the outer circle lane (close to the entry) (pcu/h);

$$k_a = \begin{cases} \frac{R_i}{R_i + LA} - \frac{L_i}{L_{imax}} & \text{per } L_i < L_{i\,max} \\ 0 & \text{in the other cases} \end{cases}$$

R_i = central island radius (m);
LA = circle width (m);
L_i = splitter island width at legs (m);

$$L_{imax} = 4{,}55 \cdot \sqrt{R_i + \frac{LA}{2}}$$

$$k_{ti} = \min \begin{cases} \frac{160}{LA \cdot (R_i + LA)} \\ 1 \end{cases} \qquad k_{te} = \min \begin{cases} 1 - \frac{(LA - 8)}{LA} \cdot \left(\frac{R_i}{(R_i + LA)}\right)^2 \\ 1 \end{cases}$$

As an example of the application of the procedure, consider a four-legged roundabout with a double-lane circle and single-lane entries, located in a rural environment. It has the following geometric features:

- external diameter: $D_{ext} = 50$ m;
- entry width: $L_e = 4.0$ m;
- splitter island width at legs: $L_i = 7.0$ m;
- circle width: $LA = 10.0$ m.

The value R_i of the central island outer radius is determined with the relationship:

$$R_i = \frac{D_{ext} - 2 \cdot LA}{2} = 15 \text{ m}$$

Since the roundabout is located in a rural area, the coefficient C_B is 3.625.

The value of parameter A, to be determined by means of Eq. (2.8), is the same for all entries and is equal to:

$$A = \frac{3600}{T_f}\left(\frac{L_e}{3.5}\right)^{0.8} = \frac{3600}{2.05}\left(\frac{4.0}{3.5}\right)^{0.8} = 1954$$

The traffic data from the TRRL procedure example (Sect. 2.1.3) are used here, and they are represented by the following matrix O/D:

$$M_{O/D} \equiv \begin{bmatrix} 0 & 150 & 300 & 200 \\ 200 & 0 & 150 & 350 \\ 350 & 150 & 0 & 150 \\ 300 & 250 & 200 & 0 \end{bmatrix}$$

$M_{O/D}$ can be used to determine the values of the flows exiting from each leg and circulating in the circle in front of each entry:

$$\begin{array}{ll} Q_{u1} = 850 \text{ pcu/h} & Q_{c1} = 600 \text{ pcu/h} \\ Q_{u2} = 550 \text{ pcu/h} & Q_{c2} = 700 \text{ pcu/h} \\ Q_{u3} = 650 \text{ pcu/h} & Q_{c3} = 750 \text{ pcu/h} \\ Q_{u4} = 700 \text{ pcu/h} & Q_{c4} = 700 \text{ pcu/h} \end{array}$$

Regarding the traffic rates Q_{ci} and Q_{ce}, we assume that about 70% of the circulating flow travels on the outer circle lane and 30% travels on the inner circle lane in front of each entry. Thus, we have the following values:

$$\begin{array}{ll} Q_{ci1} = 420 \text{ pcu/h} & Q_{ce1} = 180 \text{ pcu/h} \\ Q_{ci2} = 490 \text{ pcu/h} & Q_{ce2} = 210 \text{ pcu/h} \\ Q_{ci3} = 525 \text{ pcu/h} & Q_{ce3} = 225 \text{ pcu/h} \\ Q_{ci4} = 490 \text{ pcu/h} & Q_{ce4} = 210 \text{ pcu/h} \end{array}$$

For k_{ti} and k_{te} we calculate:

$$k_{ti} = \min\left\{\frac{160}{LA \cdot (R_i + LA)};1\right\} = \min\{0.57;\ 1\} = 0.57$$

$$k_{te} = \min\left\{1 - \frac{(LA-8)}{LA} \cdot \left(\frac{R_i}{(R_i + LA)}\right)^2;1\right\} = \min\{0.92;1\} = 0.92$$

To determine the disturbing flows in front of each leg, we must also determine the coefficient k_a that is the same for all the legs as function of L_{imax}

$$L_{imax} = 4.55 \cdot \sqrt{R_i + \tfrac{LA}{2}} = 4.55 \cdot \sqrt{15 + \tfrac{10}{2}} = 20.35 \text{ m} > L_i = 4.0 \text{ m}$$

$$k_a = \frac{R_i}{R_i + LA} - \frac{L_i}{L_{imax}} = \frac{15}{15+10} - \frac{4.0}{20.35} = 0.256$$

Therefore, with Eq. (2.9) we determine the disturbing flows in front of each leg. They are

$$\begin{aligned} Q_{d1} &= Q_{u1} \cdot k_a \cdot \left(1 - \frac{Q_{u1}}{Q_{c1}+Q_{u1}}\right) + Q_{ci1} \cdot k_{ti} + Q_{ce1} \cdot k_{te} = \\ &= 850 \cdot 0.256 \cdot \left(1 - \frac{850}{420+850}\right) + 420 \cdot 0.57 + 180 \cdot 0.92 \\ &= 526\,\text{pcu/h} \end{aligned}$$

and, similarly,

$$\begin{aligned} Q_{d2} &= 587 \text{ pcu/h} \\ Q_{d3} &= 634 \text{ pcu/h} \\ Q_{d4} &= 598 \text{ pcu/h.} \end{aligned}$$

Finally, from Eq. (2.7), the entry capacity values

$$C_1 = A \cdot e^{-C_B \cdot Q_{d1}} = 1954 \cdot e^{-3.625 \cdot 526} = 1151\text{pcu/h}$$

and, similarly,

$$\begin{aligned} C_2 &= 1082 \text{ pcu/h} \\ C_3 &= 1032 \text{ pcu/h} \\ C_4 &= 1070 \text{ pcu/h} \end{aligned}$$

Instead of applying the GIRABASE procedure, a simplified capacity formula is used (CERTU [8]) for urban French roundabouts. It is considered suitable for medium-sized and large-sized roundabouts (with a central island diameter from 20 to 60 m) with single-lane entries, symmetrical location of the legs, and a balanced entering traffic demand. In these cases, the entry capacity can be determined with the relationship:

$$C = 1500 - \frac{5}{6} \cdot Q_d \tag{2.10}$$

as function of the disturbing flow

$$Q_d = a \cdot Q_c + b \cdot Q_u \tag{2.11}$$

with

a. variable, as function of the central island radius, between 0.9 and 0.7 for $R_i <$ 15 m and > 30 m, respectively.
b. variable, as function of the splitter island width at legs, between 0 and 0.3 for L_i > 15 m and $L_i = 0$ m, respectively.

Finally, it is worthwhile to note that, in France, the GIRABASE procedure long ago replaced the other formulas developed by various French road research centers. Therefore, even the SETRA formula (See Sect. 5.2.1), which is still used extensively by Italian engineers, is no longer used in France.

2.1.5 Brilon-Wu Formula (Germany)

In Germany, based on an idea from Tanner, in 1997, Brilon and Wu proposed the following formula for the calculation of capacity C (pcu/h) of a roundabout entry (required by German Standard HBS 2001) [9]:

$$C = 3600 \cdot \left(1 - \frac{\Delta \cdot Q_c/3600}{n_c}\right)^{n_c} \cdot \frac{n_e}{T_f} \cdot \exp\left[-Q_c/3600 \cdot \left(T_c - \frac{T_f}{2} - \Delta\right)\right] \quad (2.12)$$

where:

Q_c = circulating flow in front of the entry (pcu/h);
n_c = circle lane number;
n_e = entry lane number;
T_c = critical gap;
T_f = follow-up time;
Δ = minimum headway between the vehicles circulating in the circle.

Therefore, according to Eq. (2.12), the capacity determination of roundabout entries is a function of the users' behaviors, represented by the determination of the psycho-technical times T_c, T_f, and Δ, as well as circulating traffic, the number of circle lanes, and the number of entry lanes.

Setting the model under the German conditions on the basis of experimental data (Brilon), the psycho-technical times have been estimated as $T_c = 4.1$ s, $T_f = 2.9$ s, and $\Delta = 2.1$ s.

Figure 2.8 shows the capacity behaviors determined using Eq. (2.12), as a function of the circulating traffic and of various geometric configurations of the roundabout (n_e/n_c).

Further developments of this capacity formula can be found in [9], where it is also recommended that Eq. (2.12) be used only in the case of roundabouts with a single-lane circle and single-lane entries. In the case of roundabouts that have circulatory roadways that can be used for vehicle traffic along two lines,[4] the following relationship is used to determine the capacity C (pcu/h) of an entry:

---

[4]The external diameter D_{ext} must be between 40 and 60 m, and the central line of the circle must not be marked. In addition, the semi-practicable area around the central line must not be planned. The width of the circle must be a constant value of approximately 8, with a maximum of 10 m.

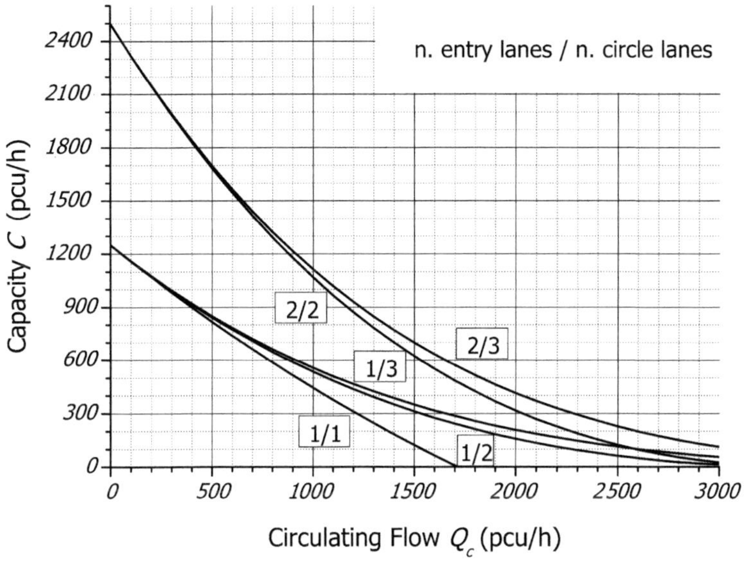

Fig. 2.8 Capacity of roundabout entries according to the Brilon-Wu formula (HBS 2001)

$$C = 3600 \cdot \frac{n_e}{T_f} \cdot \exp\left[-\frac{Q_c}{3600} \cdot \left(T_c - \frac{T_f}{2}\right)\right] \quad (2.13)$$

where:

Q_c = circulating flow in front of the entry (pcu/h);
n_e = parameter connected to the number of entry lanes; equal to 1 for single-lane entries and 1.4 for double-lane entries;
T_c = critical gap = 4.3 s;
T_f = follow-up time = 2.5 s.

As an example, we determine the entry capacity for a geometric configuration (a four-legged roundabout, with a double-lane circle and double-lane entries) and with the traffic data already presented in Sects. 2.1.1 and 2.1.2. First, we determine the circulating flows in front of each entry (Sect. 2.1.1); then we use Eq. (2.13) to determine capacity

$$\begin{aligned} C_1 &= 3600 \cdot (n_{e1}/T_f) \cdot \exp\left[-Q_{c1}/3600 \cdot (T_c - T_f/2)\right] = \\ &= 3600 \cdot (1.4/2.5) \cdot \exp\left[-376/3600 \cdot (4.3 - 2.5/2)\right] = 1194 \text{ pcu /h} \end{aligned}$$

and, similarly,

$$C_2 = 3600 \cdot (1.4/2.5) \cdot \exp\left[-379/3600 \cdot (4.3 - 2.5/2)\right] = 1191 \text{ pcu/h}$$

$$C_3 = 3600 \cdot (1.4/2.5) \cdot \exp\left[-284/3600 \cdot (4.3 - 2.5/2)\right] = 1291 \text{ pcu/h}$$

$$C_4 = 3600 \cdot (1.4/2.5) \cdot \exp\left[-276/3600 \cdot (4.3 - 2.5/2)\right] = 1299 \text{ pcu/h}$$

2.1.6 HCM 2000 Formula (USA)

The HCM 2000 approach for evaluating the entry capacities C for roundabouts is limited to schemes with one lane in the circle and one lane at the entries and with circulating flow Qc not greater than 1200 pcu/h [1]. To evaluate C, the following equation is used:

$$C = \frac{Q_c e^{-Q_c T_c/3600}}{1 - e^{-Q_c T_f/3600}} \text{ (pcu/h)} \tag{2.14}$$

where:

Q_c = circulating flow in front of the entry (pcu/h);
T_c = critical gap (s);
T_f = follow-up time (s).

Since extensive experimental data on operating roundabouts in the US were not available when the latest edition of the HCM was published, the Manual procedure gives an interval of capacity values obtained with the following values of the parameters T_c and T_f.

The upper bound of Eq. (2.14) is obtained with $T_c = 4.1$ s and $T_f = 2.6$ s, and the lower bound is obtained with $T_c = 4.6$ s and $T_f = 3.1$ s (See Fig. 2.9).

With the traffic data of Table 2.4, the capacities evaluated with Eq. (2.14) and with the German formula (2.1) were compared for a roundabout with one lane in the circle and one at the entries.

In this example the German formula is the following:

$$C = 1218 - 0.74 \cdot Q_c \text{ (pcu/h)} \tag{2.15}$$

In this case, we note that the average value between the lower and the upper bound of the capacity, evaluated by the HCM formula, is, in practice, the same as the value obtained by the German formula (2.15).

Figure 2.10 shows a comparison of two capacity formulas, the HCM capacity formula (Eq. (2.14)) and the German formula (Eq. (2.12)).

These two formulas are applied to a roundabout with single-lane entries and a single-lane circle. Equations (2.14) and (2.12) were evaluated using for T_c and T_f the values indicated by the German capacity formula ($T_c = 4.1$ s; $T_f = 2.9$ s).

We can note (See Fig. 2.9) that the American formula overestimates capacity systematically when compared to the values obtained from the German formula. This result was observed when the same geometric and traffic conditions were used,

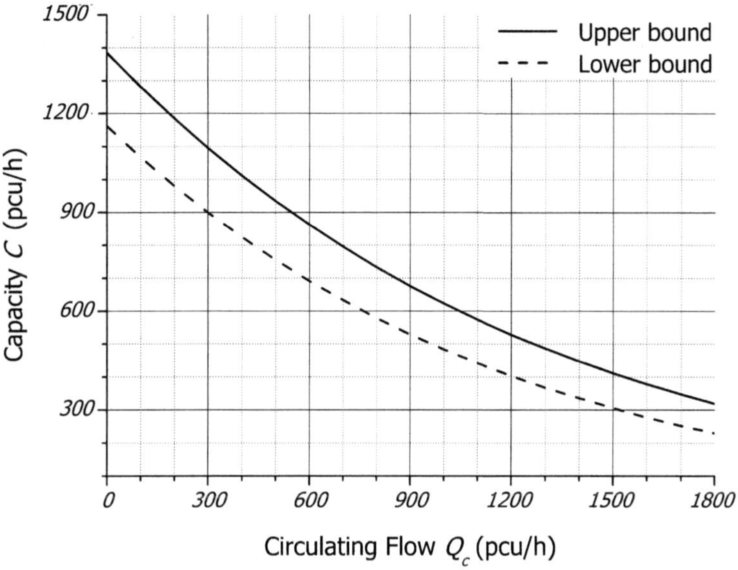

Fig. 2.9 HCM 2000 capacity formula for different values of critical gap and follow-up time

Table 2.4 Traffic data and capacity value for a roundabout

				Capacity			
				HCM 2000 formula			German formula (See Eq. 2.15)
Entry	Maneuver	Volume	Q_c	Upper bound	Lower bound	Average	
1	Right turn	22	185	1198	992	1095	1081
	Straight	208					
	Left turn	164					
2	Right turn	59	384	1023	834	929	933
	Straight	432					
	Left turn	44					
3	Right turn	187	640	834	667	751	744
	Straight	266					
	Left turn	38					
4	Right turn	40	348	1054	862	958	960
	Straight	134					
	Left turn	13					

as well as when users had the same psycho-technical parameters, for values of Q_c greater than 300–350 pcu/h.

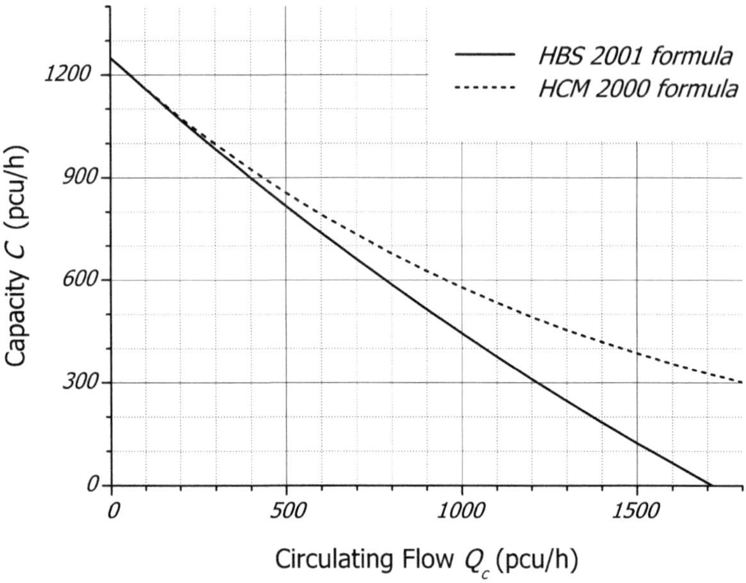

Fig. 2.10 Comparison of the German HBS 2001 capacity formula and the HCM 2000 capacity formula

2.2 Exit and Circle Capacities

Until now, no studies have been conducted that were specifically dedicated to the determination of roundabout exit and circle capacities.

Regarding exits, field observations show that the capacity limit for each lane is in the range of 1200–1400 pcu/h.[5]

Regarding the circulatory roadway, we can use the values shown in Table 2.5 concerning observations about operating roundabouts in Germany just as an indication of the values that can be expected.

2.3 Consideration of Pedestrian Crosswalks

In urban roundabouts, pedestrian crosswalks at legs reduce entry and exit capacities in proportion to the value of the pedestrian flow.

In current technical practice, three calculation procedures are mainly used to determine the above-mentioned entry capacity reductions, i.e., the English

[5] In Europe, for safety reasons, double-lane exits are rarely used.

Table 2.5 Circle capacity values according to empirical data from operating roundabouts in Germany

Type of roundabout	Number of entry lanes	Circle capacity [veh/h]
Roundabouts with single-lane circles (mini roundabouts and compact roundabouts)	1	1600
Compact roundabouts with double-lane circles	1	1600
	2	1600
Large roundabouts	1	2000
	2	2500

procedure (Marlow and Maycock), the German procedure (Brilon, Stuwe and Drews), and the French procedure (CETE de l'Ouest).

All three of the procedures are valid only when the assumption is made that pedestrians at pedestrian crosswalks have priority over vehicular traffic.

The English and French procedures are based on the same principles, since they both use the results of the mathematical queuing theory.

The German procedure is based on the treatment of empirical data obtained from operating roundabouts.

Regarding exits, specific formulas are not available at the present time, and, as will be discussed later, the same criteria that are used for exits are also used for entries.

2.3.1 Entry Capacities in Presence of Pedestrian Crosswalks

2.3.1.1 Marlow and Maycock Formula

First, the capacity value C_{ap} in the presence of only pedestrian flow Q_{ped} (Griffiths' formula) is calculated [10]

$$C_{ap} = \frac{Q_{ped}}{Q_{ped} \cdot \beta + (e^{Q_{ped} \cdot \alpha} - 1) \cdot (1 - e^{-Q_{ped}\,\beta})} \cdot 3600 \qquad (2.16)$$

where:

Q_{ped} = pedestrian flow (ped/s);
$\beta = \frac{1}{C_0}$ (s);

C_0 = capacity with pedestrian and vehicular flows equal to zero (completely empty roundabout);

$\alpha = B/v_{ped}$ = time necessary (s) to allow pedestrians to completely cross the pedestrian crosswalk, where B (m) is the width of the road at the pedestrian crosswalk, and v_{ped} (m/s) is the (mean) pedestrian flow speed.

The width B, which characterizes each entry, must be defined separately for each entry according to the roundabout geometry.

For v_{ped}, except for different direct determinations, one assumes v_{ped} = 0.5–2.0 m/s, with the suggested default value of 1.4 m/s.

Once known the value of C_{ap}, entry capacity $C_{/ped}$, which takes into account the pedestrian flow, is

$$C_{/ped} = C \cdot M \tag{2.17}$$

where M is a reduction factor of the capacity C value (veh/h) of the entry considered in absence of the pedestrian flow provided by

$$M = \frac{R^{n+2} - R}{R^{n+2} - 1} \tag{2.18}$$

with

$$R = \frac{C_{ap}}{C} \tag{2.19}$$

and "n" is equal to the number of vehicles that may be in the queuing area between the pedestrian crosswalk and the yielding line.

The term "n," which must be determined for each entry, is a function of the mean longitudinal size of the vehicles (equal to 5–6 m); to determine n, we consider all the entry lanes, e.g., in the case of a double-lane entry and a distance of 5 m between the pedestrian crosswalk and the yielding line, n = 2.

Figures 2.11, 2.12, 2.13, and 2.14 show the relationship $M = M(Q_c)$ for two urban roundabouts, i.e., a single-lane entry roundabout[6] (Figs. 2.11 and 2.12) and a double-lane entry roundabout[7] (Figs. 2.13 and 2.14). M was determined with the TRRL capacity procedure (Eq. (2.6)) and with the German HBS 2001 procedure (Brilon-Wu formula) (Eq. (2.12)). All the graphs were traced for increasing pedestrian flow values Q_{ped} from 100 to 800 ped/h with increments of 100 ped/h. The width of the pedestrian crosswalk B was assumed to be 3.5 m and 7.5 m, for single-lane entries and double-lane entries, respectively. Pedestrian speed was set to v_{ped} = 1.4 m/s.

---

[6] The geometric parameter values used are as follows: D = 34 m; u = 7 m; e = 4 m; v = 3 m; ℓ' = 7.5 m; Φ = 35°; r = 20.8 m. (The symbols for the parameter values are reported in Table 2.2.)

[7] The geometric parameter values used are as follows: D = 56 m; u = 8 m; e = 8 m; v = 6.5 m; ℓ' = 15.1 m; Φ 35°; r = 32.7 m. (The symbols for the parameter values are reported in Table 2.2).

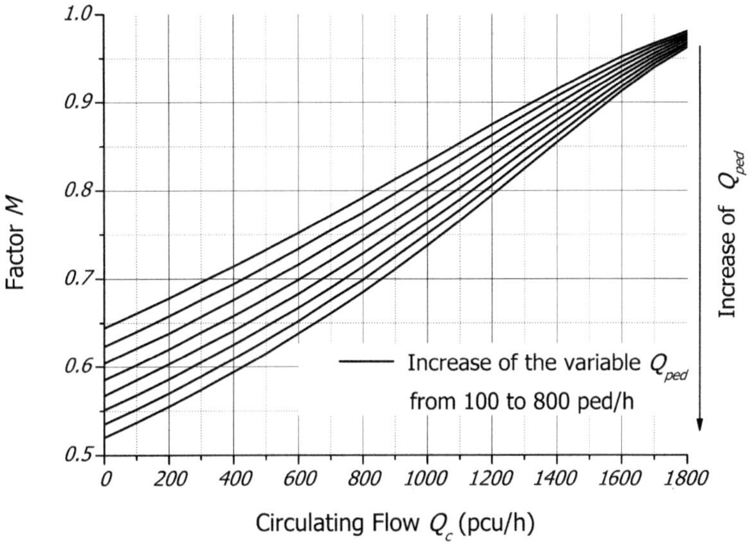

Fig. 2.11 M = M(Q_c) relationship according to Marlow and Maycock for a single-lane roundabout with single-lane entries (capacity C determined by TRRL procedure)

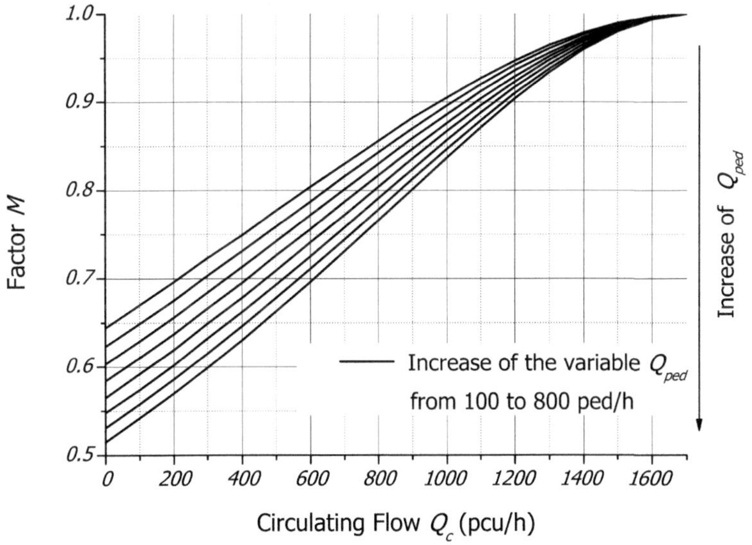

Fig. 2.12 M = M(Q_c) relationship according to Marlow and Maycock for a single-lane roundabout with single-lane entries (capacity C determined with HBS 2001 procedure)

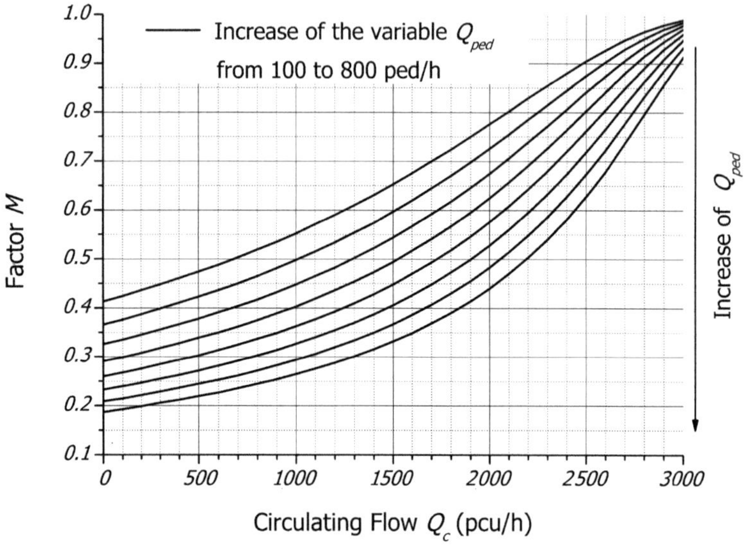

Fig. 2.13 $M = M(Q_c)$ relationship according to Marlow and Maycock for a roundabout with double-lane entries (capacity C determined with TRRL procedure)

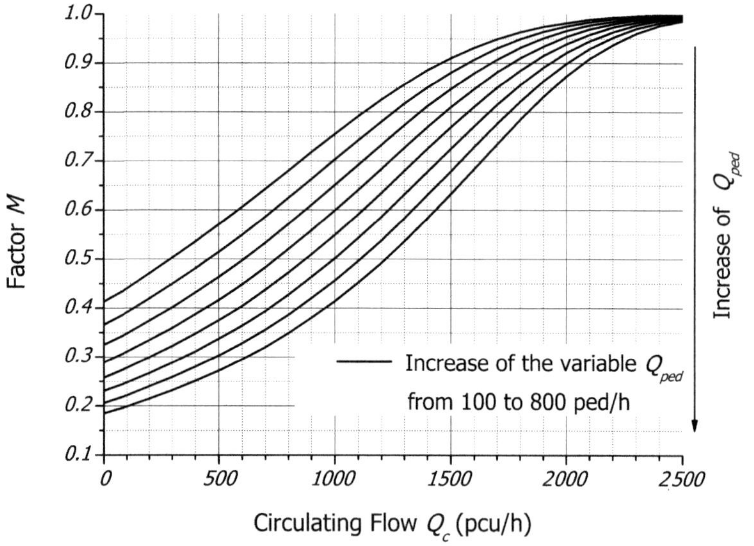

Fig. 2.14 $M = M(Q_c)$ relationship according to Marlow and Maycock for a roundabout with double-lane entries (capacity C determined with HBS 2001 procedure)

2.3.1.2 Brilon, Stuwe and Drews Formula

With this method, as with the method just illustrated in the previous section, the entry capacity C (determined with any procedure which doesn't include pedestrian crosswalks) is reduced by means of a factor M that takes into account the effects of pedestrian crosswalks [11]:

$$C_{/ped} = C \cdot M \tag{2.20}$$

M is given on the basis of entry configurations:

– *single-lane entry*

$$M = \frac{1119.5 - 0.715 \cdot Q_c - 0.644 \cdot Q_{ped} + 0.00073 \cdot Q_c \cdot Q_{ped}}{1069 - 0.65 \cdot Q_c} \tag{2.21}$$

– *double-lane entry*

$$M = \frac{1260.6 - 0.381 \cdot Q_{ped} - 0.329 \cdot Q_c}{1380 - 0.50 \cdot Q_c} \tag{2.22}$$

where

Q_c = circulating flow in front of the entry (pcu/h);
Q_{ped} = pedestrian flow crossing the leg (ped/h).

Figures 2.15 and 2.16 show the relationship $M = M(Q_c)$ for a single-lane entry roundabout (See Fig. 2.15) and a double-lane entry roundabout (See Fig. 2.16). The graphs were traced for increasing pedestrian flow values of Q_{ped} from 100 to 800 ped/h with increments of 100 ped/h.

The equations that allow the determination of the reduction factor M may give unrealistic results, when used outside the existence intervals of the experimental measures. Thus, for example, in the case of single-lane entry roundabouts with a small volume of pedestrians (<100 ped/h), the formulas show that when there is a marginal increase in pedestrians Q_{ped}, capacity also tends to increase.

However, these circumstances do not invalidate the formula, but they demand careful application.

2.3.1.3 CETE de l'Ouest Formula

Also with this procedure, entry capacity C is reduced by means of a factor F that takes into account the pedestrian flow [5]

$$C_{/ped} = C \cdot F \tag{2.23}$$

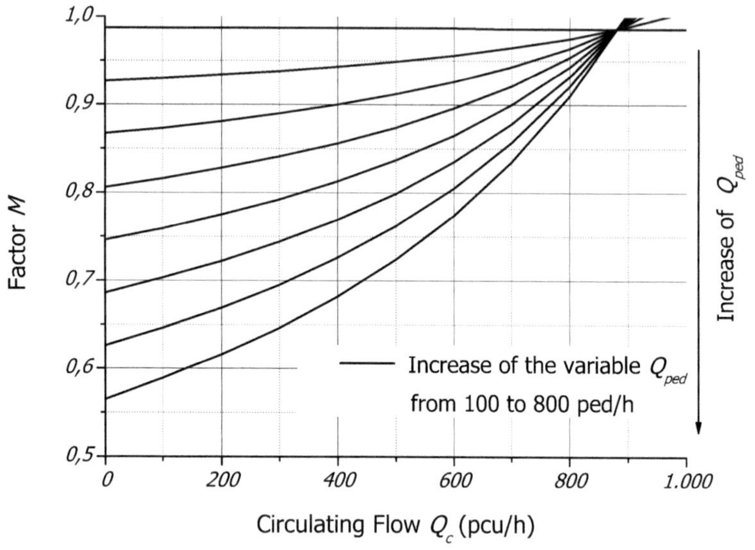

Fig. 2.15 $M = M(Q_c)$ relationship according to Brilon, Stuwe and Drews for a single-lane entry roundabout

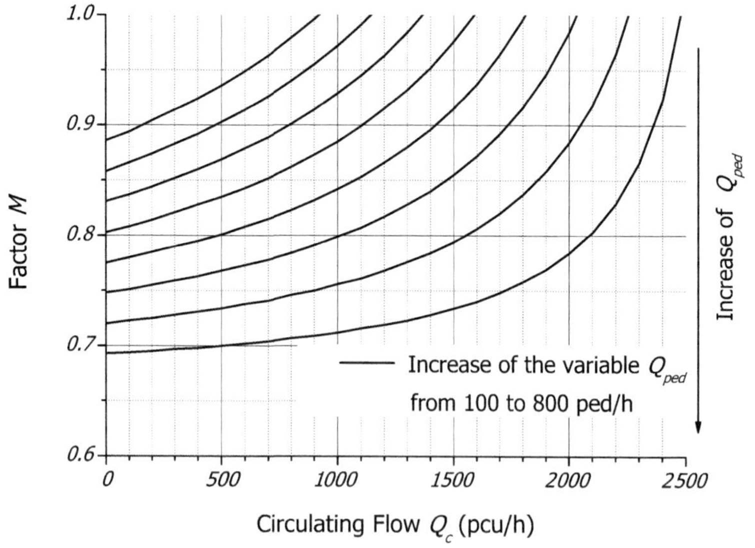

Fig. 2.16 $M = M(Q_c)$ relationship according to Brilon, Stuwe and Drews for a double-lane entry roundabout

with

$$F = 1 - \exp(-k \cdot Q_d \cdot \beta) \cdot [1 - \exp(-Q_{ped} \cdot T)] \quad (2.24)$$

where:

Q_d = disturbing traffic in front of the entry (pcu/s) (Q_d must be determined according to the capacity formula chosen, e.g., Q_d is given by Eq. (2.9) if one uses the GIRABASE procedure);
Q_{ped} = pedestrian flow crossing the leg (ped/s);

$\beta = \frac{1}{C_0}$ (s);

C_0 = capacity with pedestrian and vehicular flows equal to zero (completely empty roundabout);
k = number of vehicles that may be in the area between the pedestrian crosswalk and the yielding line.

The previous graphs were traced for increasing pedestrian flow values Q_{ped} from 100 to 800 ped/h with increments of 100 ped/h.

Also Figure 2.17 shows the relationship $F = F(Q_c)$ for increasing pedestrian flow values Q_{ped} from 100 to 800 ped/h with increments of 100 ped/h.

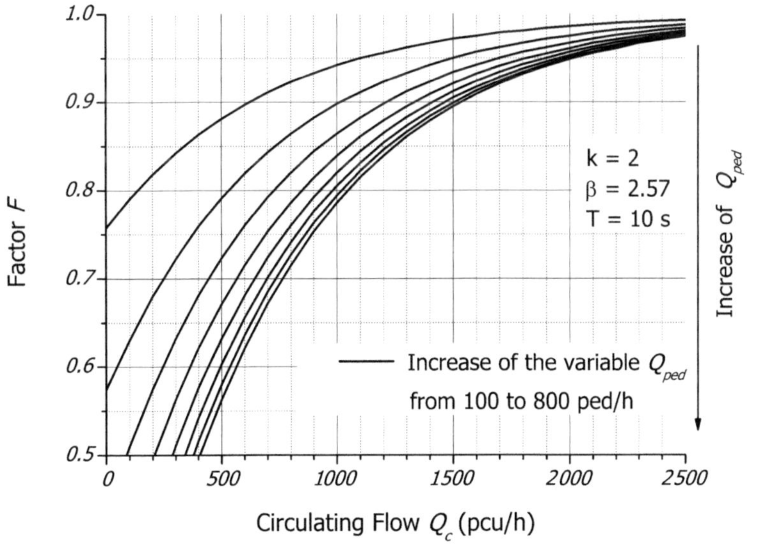

Fig. 2.17 $F = F(Q_c)$ relationship according to the CETE formula

2.3.2 *Exit Capacities in Presence of Pedestrian Crosswalks*

We can reasonably assume that a heavy pedestrian flow at an exit causes a capacity reduction of the exit. At first, these effects are generally and approximately determined with the formula by Marlow and Maycock (Sect. 2.3.1.1), even though recent studies have shown that, using the equations proposed by this method, the pedestrian influence on the roundabout exit capacity is overestimated.

In conclusion, we wish to emphasize that the formulas relative to the determination of roundabout exit capacities demand careful application because they have not been specifically validated at this time.

2.4 Some Concluding Remarks on Capacity Formulations

The differences between the formulas presented in the previous sections and between these formulas and other formulas from the literature are mainly caused by the following reasons, as are the discrepancies between their capacity estimates:

- drivers' behaviors at the intersection are due, among other things, to the extent of their experience with roundabouts on the road networks of their countries;
- roundabout geometric standards vary from nation to nation, which makes them dissimilar from configurations that have the same number of circle lanes and entry lanes;
- for the mixed traffic data used to determine and set calculation procedures. Some of these data are specific for certain environments, such as urban areas and rural areas, whereas other data are from contexts that are only generically similar to the environments from which the most significant samples come;
- the environments where the roundabouts are built are different because of varying national urban and territorial features, even though different environments are called in the same way in the different languages;
- there are correlations between the geometric variables and traffic values that can highlight or hide the role of some parameters during the experimental phase and statistical treatment of measures. Thus, these parameters may not be present in the capacity formulas.

In some countries (as in Italy and Spain), a specific formula for the determination of circular intersection capacity has not yet been defined.

Therefore, in these countries, it is essential to compare the different types of roundabouts present in the national standard (if one is available) and in foreign standards that address the same types of intersections. Thus, we can better choose the best calculation procedure for the case under examination, at least with respect to the conformity between the geometric standards and the environment.

As we have just mentioned, different nations use different classifications and nomenclature, resulting in their standards not being fully compatible or interchangeable. (See, for example, Tables 2.6 and 2.7, which compare Italian, German, and Swiss definitions).

Table 2.6 Roundabout classification: Italian versus German nomenclatures

	Italian nomenclatures	German nomenclatures	
	D_{ext} (m)	D_{ext} (m)	
Mini-roundabouts	14–25	13–24	
Compact roundabouts	25–40	26–60	urban: 26–35 (single-lane circle) rural: 30–45 (double-lane circle) Urban and rural: 40–60 (double-lane circle)
Roundabouts	40–50	55–80	
"Rotary circulation" layout	>50	–	

Table 2.7 Roundabout classification: Italian versus Swiss nomenclatures

	Italian nomenclatures	Swiss nomenclatures
	D_{ext} (m)	D_{ext} (m)
Mini-roundabouts	14–25	14–20 (town centers, residential areas, urban areas)
Small roundabouts	–	19–25 (town centers, residential urban, and suburban areas)
Compact roundabouts	25–40	25–35 (urban, suburban, and rural areas)
Roundabouts	40–50	–
Big roundabouts (Swiss Standard nomenclature)	–	>35 (rural areas)
"Rotary circulation" layout (Italian Standard nomenclature)	>50	–

On the other hand, one must make such comparisons in order to decide which capacity relationship to use.

In fact, the users' behaviors are not comparable when specific investigations are not available.

Finally, for engineers of a country where a capacity formulation is not available, the selection of a capacity formula is not easy and must be seriously considered.

However, when the engineers analyze the results obtained, they can use the comparison with the results derived from a different formula that has been deemed to be suitable for the case in question and/or compare the results with the ones of a traffic micro-simulation study of the roundabout.

2.5 Capacity Calculation at Saturation or Oversaturation Conditions of Entries

If traffic demand at one or more entries equals or exceeds capacity, the system reaches a state characterized by flows coming from undersaturated entries equal to demand and from saturated or oversaturated entries equal to capacity.

Now we recall that, an entry capacity depends on the circulating flows determined by entering flows coming from the other legs of the roundabout. Therefore, the entry capacity values calculated only on the basis of traffic demand at the intersection and without taking into account the saturation or oversaturation conditions of some entries are not correct.

Thus, we will now present an iterative method for the determination of entering flows to a roundabout and the determination of the respective capacity values for saturated or oversaturated conditions at one or more entries.

This procedure is illustrated for a four-legged roundabout, but it is always valid due to the mathematical nature of the problem.

Traffic demand is known and is represented by the flow vectors $[Q_{ei}]$ and matrix $P_{O/D}$, i.e., matrix $M_{O/D}$ (See Sect. 1.1).

We proceed with successive calculation steps; for each step (k), we determine the flow values for each leg that can actually enter the roundabout. Thus, we obtain a vector $[Q_{ei}^{*(k)}]$; the iterative calculation converges very rapidly and ends when the same entering flows are reached for two successive steps, i.e., $[Q_{ei}^{*(k-1)}] = [Q_{ei}^{*(k)}]$.

The values represent the volumes that can enter the roundabout examined, under the restraint that one or more legs must have demands equal to or exceeding their capacities (traffic rate $\rho_i = Q_{ei}/C_i \geq 1$, i.e., reserve capacity $RC_i = C_i - Q_{ei} \leq 0$).

At step (1) of the procedure, we assume that the disturbing traffic is zero in front of entry 1 ($Q_{d1}^{(1)} = 0$). Then, we determine the corresponding capacity $C_1^{(1)}$ value and the flow $Q_{e1}^{*(1)}$ as the smaller of the calculated capacity value and traffic demand Q_{e1}.

For entry 2, on the basis of flow $Q_{e1}^{*(1)}$ and traffic percentage matrix $P_{O/D}$, we determine the disturbing traffic value ($Q_{d2}^{(1)}$), the corresponding capacity $C_2^{(1)}$

value, and the flow $Q_{e2}^{*(1)}$, as the smaller of the calculated capacity value and traffic demand Q_{e2}.

We proceed similarly for entries 3 and 4. Thus, for entry 3, we determine the disturbing traffic ($Q_{d3}^{(1)}$) (starting with $Q_{e1}^{*(1)}$, $Q_{e2}^{*(1)}$, and the traffic percentage matrix), capacity $C_3^{(1)}$ and the flow $Q_{e3}^{*(1)}$; for entry 4, the disturbing traffic ($Q_{d4}^{(1)}$) (starting with $Q_{e1}^{*(1)}$, $Q_{e2}^{*(1)}$, $Q_{e3}^{*(1)}$ and the traffic percentage matrix), capacity $C_4^{(1)}$ and the flow $Q_{e4}^{*(1)}$. In the end, we obtain the vector $[Q_{ei}^{*(1)}]$ of the entering flows at step (1).

At step (2), we repeat the calculations starting with entry 1, and we determine the disturbing traffic ($Q_{d1}^{(2)}$) (starting with $Q_{e2}^{*(1)}$, $Q_{e3}^{*(1)}$, $Q_{e4}^{*(1)}$ and traffic percentage matrix), capacity $C_1^{(2)}$, and the flow $Q_{e1}^{*(2)}$.

We proceed similarly, using an iterative method, for all of the other entries, until we obtain the vector $[Q_{ei}^{*(2)}]$ of the entering flows at step (2).

Regarding, in particular, the determination of disturbing flow $Q_{di}^{(k)}$ for the generic leg "i" at step (k) of the iterative method, it is worth emphasizing that it must be determined by means of the "most recent" values of the volumes that may enter the roundabout from the other entry; thus, for example, for $Q_{d2}^{(2)}$, one uses the flow values $Q_{e1}^{*(2)}$, $Q_{e3}^{*(1)}$, and $Q_{e4}^{*(1)}$; for $Q_{d3}^{(2)}$, one uses, $Q_{e1}^{*(2)}$, $Q_{e2}^{*(2)}$, and $Q_{e4}^{*(1)}$, and so the process continues until it is completed.

We proceed similarly, using an iterative method, for the following calculation steps. Thus, we obtain a succession of vectors of entering flows $[Q_{ei}^{*(3)}]$, ..., $[Q_{ei}^{*(k)}]$; as already discussed, the calculation converges very rapidly and ends when the same entering volumes ($[Q_{ei}^{*(k-1)}] = [Q_{ei}^{*k}]$) occur for two successive steps and the same disturbing flows ($[Q_{di}^{*(k-1)}] = [Q_{di}^{*k}]$) and capacities $[C_i^{*(k-1)} = C_i^{*k}]$ are reached.

Thus, we have determined the respective capacity values taking into account the entering flow values at the intersection and the saturation or oversaturation conditions of some of the entries. In particular, it is worth noting that, for the saturated or oversaturated legs, the capacity values C_i evidently coincide with those of the volumes Q_{ei} that may enter the roundabout.

2.5.1 A Worked Example

We will now present an application of the procedure just illustrated. For the sake of simplicity, we use a capacity formula for which the disturbing traffic Q_d consists only of the circulating flow Q_c in front of the entries ($Q_d = Q_c$). However, the procedure can be used without difficulty, even when Q_d is also function of the exiting flow of the leg. (See, for example, Eq. (2.9)).

Consider a four-legged roundabout for which the "traffic percentage matrix" is:

$$P_{O/D} = \begin{bmatrix} 0.00 & 0.31 & 0.38 & 0.31 \\ 0.24 & 0.00 & 0.44 & 0.32 \\ 0.36 & 0.40 & 0.00 & 0.24 \\ 0.30 & 0.30 & 0.40 & 0.00 \end{bmatrix}$$

The traffic demand vector (pcu/h) is

$$[\mathrm{Q_{ei}}] = \begin{bmatrix} 800 & 500 & 900 & 700 \end{bmatrix}$$

Therefore, the matrix O/D is

$$[\mathrm{O/D}] = \begin{bmatrix} 0 & 248 & 304 & 248 \\ 120 & 0 & 220 & 160 \\ 324 & 360 & 0 & 216 \\ 210 & 210 & 280 & 0 \end{bmatrix}$$

Among the roundabout capacity formulas available in the literature, we use the Brilon-Bondzio formula (2.1), as an example,

$$\mathrm{C} = \mathrm{A} - \mathrm{B} \cdot \mathrm{Q_c}$$

which is specialized for roundabouts with a single-lane circle and single-lane legs ($\mathrm{A} = 1218$, $\mathrm{B} = 0.74$).

Starting with traffic data, we determine the circulating flow vector $[\mathrm{Q_{ci}}]$ (pcu/h) (Eq. (1.12) in Chap. 1).

$$[\mathrm{Q_{ci}}] = \begin{bmatrix} 850 & 832 & 528 & 804 \end{bmatrix}$$

The capacity vector (pcu/h), determined through the application of the above-mentioned formula for each leg, is

$$[\mathrm{C_i}] = \begin{bmatrix} 589 & 602 & 827 & 623 \end{bmatrix}$$

Comparing the traffic demand vector $[\mathrm{Q_{ei}}]$ to the capacity vector $[\mathrm{C_i}]$, we can notice that entries 1, 3, and 4 are at oversaturation conditions ($[\rho_i] = [\mathrm{Q_i}/\mathrm{C_i}] =$ [1.36 0.83 1.09 1.12]).

As already discussed, the circulating flow values and, therefore, the capacity values just determined are not correct under such conditions. In fact, the actual flow entering the oversaturated legs is not equal to demand, but to capacity.

Therefore, if one or more entries are saturated, we use the procedure calculation to determine a balanced situation.

For entry 1, if we assume at step (1) that $\mathrm{Q_{c1}^{(1)}} = 0$, then we obtain $\mathrm{C_1^{(1)}} = 1218 - 0.74 \cdot \mathrm{Q_{c1}^{(1)}} = 1218 - 0.74 \cdot 0 = 1218$ pcu/h and $\mathrm{Q_{e1}^{*(1)}} = \min(\mathrm{C_1^{(1)}}; \mathrm{Q_{e1}}) = \min(1218; 800) = 800$ pcu/h.

This volume is divided into the following, on the basis of the traffic percentage matrix:

$$
\begin{aligned}
Q_{12}^{*(1)} &= P_{12} \cdot Q_{e1}^{*(1)} = 0.31 \cdot 800 = 248\,\text{pcu/h} \\
Q_{13}^{*(1)} &= P_{13} \cdot Q_{e1}^{*(1)} = 0.38 \cdot 800 = 304\,\text{pcu/h} \\
Q_{14}^{*(1)} &= P_{14} \cdot Q_{e1}^{*(1)} = 0.31 \cdot 800 = 248\,\text{pcu/h}
\end{aligned}
$$

For entry 2, we determine

$$
\begin{aligned}
Q_{c2}^{(1)} &= Q_{13}^{*(1)} + Q_{14}^{*(1)} = 304 + 248 = 552\,\text{pcu/h} \\
C_{2}^{(1)} &= 1218 - 0.74 \cdot Q_{c2}^{(1)} = 1218 - 0.74 \cdot 552 = 810\,\text{pcu/h} \\
Q_{e2}^{*(1)} &= \min(C_{2}^{(1)}, Q_{e2}) = \min(810, 500) = 500\,\text{pcu/h}
\end{aligned}
$$

This volume is divided into the following, on the basis of the traffic percentage matrix:

$$
\begin{aligned}
Q_{21}^{*(1)} &= P_{21} \cdot Q_{e2}^{*(1)} = 0.24 \cdot 500 = 120\,\text{pcu/h} \\
Q_{23}^{*(1)} &= P_{23} \cdot Q_{e2}^{*(1)} = 0.44 \cdot 500 = 220\,\text{pcu/h} \\
Q_{24}^{*(1)} &= P_{24} \cdot Q_{e2}^{*(1)} = 0.32 \cdot 500 = 160\,\text{pcu/h}
\end{aligned}
$$

For entry 3, we determine

$$
\begin{aligned}
Q_{c3}^{(1)} &= Q_{14}^{*(1)} + Q_{21}^{*(1)} + Q_{24}^{*(1)} = 248 + 120 + 160 = 528\,\text{pcu/h} \\
C_{3}^{(1)} &= 1218 - 0.74 \cdot Q_{c3}^{(1)} = 1218 - 0.74 \cdot 528 = 827\,\text{pcu/h} \\
Q_{e3}^{*(1)} &= \min(C_{3}^{(1)}, Q_{e3}) = \min(827; 900) = 827\,\text{pcu/h}
\end{aligned}
$$

This volume is divided into:

$$
\begin{aligned}
Q_{31}^{*(1)} &= P_{31} \cdot Q_{e3}^{*(1)} = 0.36 \cdot 827 = 298\,\text{pcu/h} \\
Q_{32}^{*(1)} &= P_{32} \cdot Q_{e3}^{*(1)} = 0.40 \cdot 827 = 331\,\text{pcu/h} \\
Q_{34}^{*(1)} &= P_{34} \cdot Q_{e3}^{*(1)} = 0.24 \cdot 827 = 198\,\text{pcu/h}
\end{aligned}
$$

For entry 4, we determine

$$
\begin{aligned}
Q_{c4}^{(1)} &= Q_{21}^{*(1)} + Q_{31}^{*(1)} + Q_{32}^{*(1)} = 120 + 298 + 331 = 749\,\text{pcu/h} \\
C_{4}^{(1)} &= 1218 - 0.74 \cdot Q_{c4}^{(1)} = 1218 - 0.74 \cdot 749 = 664\,\text{pcu/h} \\
Q_{e4}^{(1)} &= \min(C_{4}^{(1)}; Q_{e4}) = \min(664; 700) = 664\,\text{pcu/h}
\end{aligned}
$$

This volume is divided into:

$$
\begin{aligned}
Q_{41}^{*(1)} &= P_{41} \cdot Q_{e4}^{*(1)} = 0.30 \cdot 664 = 199\,\text{pcu/h} \\
Q_{42}^{*(1)} &= P_{42} \cdot Q_{e4}^{*(1)} = 0.30 \cdot 664 = 199\,\text{pcu/h} \\
Q_{43}^{*(1)} &= P_{43} \cdot Q_{e4}^{*(1)} = 0.40 \cdot 664 = 266\,\text{pcu/h}
\end{aligned}
$$

Table 2.8 Circulating flows [Q_{ci}], capacity [C_i], entering flows [Q^*_{ei}], and matrix [$M^*_{O/D}$] at step (1) of the iterative method

Leg	[$Q_{ci}^{(1)}$]	[$C_i^{(1)}$]	[$Q_{ei}^{*(1)}$]	O \ D	1	2	3	4
1	0	1218	800	1	0	248	304	248
2	552	810	500	2	120	0	220	160
3	528	827	827	3	298	331	0	198
4	749	664	664	4	199	199	266	0

For step (1), Table 2.8 shows the circulating flow values [Q_{ci}], capacity values [C_i], entering flow values [$Q_{ei}{}^*$], and matrix $M^*_{O/D}$.

The calculations proceed in an iterative way; in particular, it is worth noting that, as already discussed, the circulating flow $Q_{ci}{}^{(k)}$ in front of the generic leg "i" at step (k) of the procedure must be determined by means of the "most updated" values of the volumes that may enter the roundabout from the other legs, based on the previous calculation steps.

For entry 1, we obtain

$$
\begin{aligned}
Q_{c1}^{(2)} &= Q_{42}^{*(1)} + Q_{43}^{*(1)} + Q_{32}^{*(1)} = 199 + 266 + 331 = 796\,\text{pcu/h}\\
C_1^{(2)} &= 1218 - 0.74 \cdot Q_{c1}^{(2)} = 1218 - 0.74 \cdot 796 = 629\,\text{pcu/h}\\
Q_{e1}^{*(2)} &= \min(C_1^{(2)}, Q_{e1}) = \min(629, 800) = 629\,\text{pcu/h}
\end{aligned}
$$

and therefore

$$
\begin{aligned}
Q_{12}^{*(2)} &= P_{12} \cdot Q_{e1}^{*(2)} = 0.31 \cdot 629 = 195\,\text{pcu/h}\\
Q_{13}^{*(2)} &= P_{13} \cdot Q_{e1}^{*(2)} = 0.38 \cdot 629 = 239\,\text{pcu/h}\\
Q_{14}^{*(2)} &= P_{14} \cdot Q_{e1}^{*(2)} = 0.31 \cdot 629 = 195\,\text{pcu/h}
\end{aligned}
$$

Similarly, for entry 2 we determine

$$
\begin{aligned}
Q_{c2}^{(2)} &= Q_{13}^{*(2)} + Q_{14}^{*(2)} + Q_{43}^{*(1)} = 239 + 195 + 266 = 700\,\text{pcu/h}\\
C_2^{(2)} &= 1218 - 0.74 \cdot Q_{c2}^{(2)} = 1218 - 0.74 \cdot 700 = 700\,\text{pcu/h}\\
Q_{e2}^{*(2)} &= \min(C_2^{(2)}, Q_{e2}) = \min(700; 500) = 500\,\text{pcu/h}
\end{aligned}
$$

and therefore

$$
\begin{aligned}
Q_{21}^{*(2)} &= P_{21} \cdot Q_{e2}^{*(2)} = 0.24 \cdot 500 = 120\,\text{pcu/h}\\
Q_{23}^{*(2)} &= P_{23} \cdot Q_{e2}^{*(2)} = 0.44 \cdot 500 = 220\,\text{pcu/h}\\
Q_{24}^{*(2)} &= P_{24} \cdot Q_{e2}^{*(2)} = 0.32 \cdot 500 = 160\,\text{pcu/h}
\end{aligned}
$$

For entry 3, we have

$$
\begin{aligned}
Q_{c3}^{(2)} &= Q_{14}^{*(2)} + Q_{21}^{*(2)} + Q_{24}^{*(2)} = 195 + 120 + 160 = 475\,\text{pcu/h} \\
C_{3}^{(2)} &= 1218 - 0.74 \cdot Q_{c3}^{(2)} = 1218 - 0.74 \cdot 475 = 867\,\text{pcu/h} \\
Q_{e3}^{*(2)} &= \min(C_{3}^{(2)}, Q_{e3}) = \min(867, 900) = 867\,\text{pcu/h}
\end{aligned}
$$

and therefore

$$
\begin{aligned}
Q_{31}^{*(2)} &= P_{31} \cdot Q_{e3}^{*(2)} = 0.36 \cdot 867 = 312\,\text{pcu/h} \\
Q_{32}^{*(2)} &= P_{32} \cdot Q_{e3}^{*(2)} = 0.40 \cdot 867 = 347\,\text{pcu/h} \\
Q_{34}^{*(2)} &= P_{34} \cdot Q_{e3}^{*(2)} = 0.24 \cdot 867 = 208\,\text{pcu/h}
\end{aligned}
$$

Finally, we determine for entry 4

$$
\begin{aligned}
Q_{c4}^{(2)} &= Q_{21}^{*(2)} + Q_{31}^{*(2)} + Q_{32}^{*(2)} = 120 + 312 + 346 = 778\,\text{pcu/h} \\
C_{4}^{(2)} &= 1218 - 0.74 \cdot Q_{c4}^{(2)} = 1218 - 0.74 \cdot 778 = 642\,\text{pcu/h} \\
Q_{e4}^{(2)} &= \min(C_{4}^{(2)}; Q_{e4}) = \min(642; 700) = 642\,\text{pcu/h}
\end{aligned}
$$

and therefore

$$
\begin{aligned}
Q_{41}^{*(2)} &= P_{41} \cdot Q_{e4}^{*(2)} = 0.30 \cdot 642 = 193\,\text{pcu/h} \\
Q_{42}^{*(2)} &= P_{42} \cdot Q_{e4}^{*(2)} = 0.30 \cdot 642 = 193\,\text{pcu/h} \\
Q_{43}^{*(2)} &= P_{43} \cdot Q_{e4}^{*(2)} = 0.40 \cdot 642 = 257\,\text{pcu/h}
\end{aligned}
$$

For step (2), Table 2.9 shows the circulating flow values $[Q_{ci}]$, capacity values $[C_i]$, entering flow values $[Q_{ei}{}^*]$, and matrix $M_{O/D}^*$.

Proceeding similarly for the successive calculation phases, we obtain the values shown in Tables 2.10 and 2.11 (steps (3) and (4)).

Table 2.9 Circulating flows $[Q_{ci}]$, capacity $[C_i]$, entering flows $[Q_{ei}^*]$, and matrix $M_{O/D}^*$ at step (2) of the iterative method

Leg	$[Q_{ci}^{(2)}]$	$[C_i^{(2)}]$	$[Q_{ei}^{*(2)}]$	O \ D	1	2	3	4
1	796	629	629	1	0	195	239	195
2	700	700	500	2	120	0	220	160
3	475	867	867	3	312	346	0	208
4	778	642	642	4	193	193	256	0

Table 2.10 Circulating flows $[Q_{ci}]$, capacity $[C_i]$, entering flows $[Q^*_{ei}]$, and matrix $M^*_{O/D}$ at step (3) of the iterative method

Leg	$[Q^{(3)}_{ci}]$	$[C^{(3)}_{i}]$	$[Q^{*(3)}_{ei}]$	O \ D	1	2	3	4
1	796	629	629	1	0	195	239	195
2	691	707	500	2	120	0	220	160
3	475	867	867	3	312	347	0	208
4	778	642	642	4	193	193	256	0

Table 2.11 Circulating flows $[Q_{ci}]$, capacity $[C_i]$, entering flows $[Q^*_{ei}]$, and matrix $M^*_{O/D}$ at step (4) of the iterative method

Leg	$[Q^{(4)}_{ci}]$	$[C^{(4)}_{i}]$	$[Q^{*(4)}_{ei}]$	O \ D	1	2	3	4
1	796	629	629	1	0	195	239	195
2	691	707	500	2	120	0	220	160
3	475	867	867	3	312	347	0	208
4	778	642	642	4	193	193	256	0

At calculation steps (3) and (4), the entering flow vectors $[Q^*_{ei}]$, circulating flow vectors $[Q^*_{ci}]$, and capacity vectors $[C_i]$ coincide.[8]

The iterative process converges, i.e., $[Q^*_{ei}] = [629\ 500\ 867\ 642]$ is the flow vector that enters the roundabout taking into account the oversaturation of entries 1, 3, and 4; $[C^*_i] = [629\ 707\ 867\ 642]$ is the capacity vector.

As already pointed out, we can notice that, for oversaturated entries, the flow values that may enter the roundabout are equal to the capacity values of the respective entries. On the other hand, for entry 2, which is at undersaturation conditions, the entering flow value is equal to traffic demand.

2.6 Calculation Procedures for Simple Capacity and Total Capacity

As already pointed out in Sect. 1.2, roundabout performance indicators as a whole are, in technical practice, "simple capacity" SC and "whole capacity" or "total capacity" TC.

The determinations of SC and TC are not straightforward, and they require the development of suitable computational procedures.

[8] The iterative results and the final results are all given approximated integers.

For the sake of simplicity, but without narrowing our general aims, we will now present these procedures for a four-legged roundabout with an assigned traffic demand (initial condition) represented by the flow vector $[Q_{ei}]$ and by matrix $P_{O/D}$ (i.e., matrix $M_{O/D}$).

To calculate simple capacity means to determine a specific condition that is characterized by equivalence between demand and capacity values for at least one of the entries. This is done by means of a uniform increase in entering flow from an initial traffic state.

1) we calculate the disturbing flows (Q_{di}) for each entry starting with the demand vector $[Q_{ei}]$ and traffic percentage matrix $P_{O/D}$;
2) for each entry, we express the entering flow and disturbing flow values under saturation conditions as the multiplication of traffic demand by a suitable multiplier (different for each entry), i.e., $Q^*_{ei} = \delta_i Q_{ei}$ and $Q^*_{di} = \delta_i Q_{di}$.
   Thus, the generic capacity formula used for the calculation can be written as $\delta_i Q_{ei} = f_i(\delta_i Q_{di})$. Solving these equations, we obtain the values of the multipliers δ_i (one for each entry) of traffic demands at legs when these same legs are saturated;
3) we multiply demand vector $[Q_{ei}]$ by the smallest multiplier that we have found (corresponding to the first one that reaches saturation), and, thus, we determine simple capacity SC.

To determine total capacity, we use the following iterative method:

1) we assume an arbitrary, first-attempt, entering flow vector $[Q^{(1)}_{ei}] = [Q^{(1)}_{e1}\ Q^{(1)}_{e2}\ Q^{(1)}_{e3}\ Q^{(1)}_{e4}]$;
2) for entry 1 we determine the disturbing traffic $Q^{(1)}_{d1}$ as a function of the first attempt flow vector $[Q^{(1)}_{ei}]$ and matrix $[P_{ij}]$, and, by the capacity formula used, we determine a new value of entering flow 1 $Q^{(2)}_{e1}$;
3) for entries 2, 3, and 4, we proceed similarly. First, we determine the disturbing flows (using the most "updated" entering flows for each leg, i.e., those obtained with the application of the iterative method). Then, by means of the capacity formula, we determine the new entering flow values, obtaining for them a second vector $[Q^{(2)}_{ei}]$;
4) we repeat the calculation process until we obtain, for two successive steps, two entering flow vectors that are equal, i.e., when $[Q^{(k-1)}_{ei}] = [Q^{(k)}_{ei}]$.

The entering flow vector thus determined represents the number of the entering vehicles from each leg when all the entries are simultaneously saturated. The roundabout total capacity value TC is the sum of all these flows.

Finally, it is worth noting that total capacity is a function only of the traffic percentage matrix $P_{O/D}$, and, therefore, it can be determined starting with an arbitrary demand vector. In other words, with equal percentage distribution, total capacity vector is univocal, independently of the demand vector used for the first calculation step (first-attempt demand vector). Contrary to total capacity, it should be noted that simple capacity is function of both traffic demand and its percentage distribution (vector $[Q_{ei}]$ and matrix $P_{O/D}$).

2.6.1 A Worked Example

We will now present a example calculation of simple capacity and total capacity.

Consider a four-legged roundabout with a traffic percentage matrix $P_{O/D}$ of:

$$P_{O/D} = \begin{bmatrix} 0.00 & 0.15 & 0.75 & 0.10 \\ 0.19 & 0.00 & 0.24 & 0.57 \\ 0.63 & 0.15 & 0.00 & 0.22 \\ 0.19 & 0.74 & 0.07 & 0.00 \end{bmatrix}$$

We use the Brilon-Bondzio (2.1) formula as the capacity formula.

$$C = A - B \cdot Q_c$$

This formula is specialized for roundabouts with a single-lane circle and single-lane legs ($A = 1218$, $B = 0.74$).

Traffic demand vector (pcu/h) is equal to

$$[Q_{ei}] = [160\ 100\ 240\ 200]$$

2.6.1.1 Simple Capacity Calculation

To calculate simple capacity, we first determine the circulating flow vector $[Q_{ci}]$ (pcu/h). (Eq. (1.12) in Chap. 1).

$$[Q_{ci}] = \left[\ 198\ 150\ 92\ 206\ \right]$$

For each entry, we look for the initial flow multiplier that causes saturation at the entry by means of the relationship $\delta_i Q_{ei} = A–B \cdot \delta_i Q_{ci}$, which, when applied to the case under examination, becomes

$$\begin{aligned} \delta_1 \cdot 160 &= 1218 - 0.74 \cdot \delta_1 \cdot 198 \\ \delta_2 \cdot 100 &= 1218 - 0.74 \cdot \delta_2 \cdot 150 \\ \delta_3 \cdot 240 &= 1218 - 0.74 \cdot \delta_3 \cdot 92 \\ \delta_4 \cdot 200 &= 1218 - 0.74 \cdot \delta_4 \cdot 206 \end{aligned}$$

from which we have

$$\delta_1 = 3.97$$
$$\delta_2 = 5.77$$
$$\delta_3 = 3.95$$
$$\delta_4 = 3.46$$

Multiplying traffic demand $[Q_{ei}]$ by the smaller multiplier ($\delta_4 = 3.46$), we determine the entering flow values that correspond to the first roundabout congestion event for a uniform increase in the flows (i.e., simple capacity) $[Q_{ei}^{(SC)}] = [SC]$ (pcu/h):

$$[Q_{ei}^{(SC)}] = [SC] = [553\ 346\ 829\ 691]$$

Under traffic conditions for which entering flow values are equal to simple capacity, we can also determine the circulating flow values (pcu/h) in front of the entries.

$$[Q_{ci}^{(SC)}] = [684\ 518\ 318\ 712]$$

capacities (pcu/h)

$$[C^{(SC)}] = [712\ 834\ 983\ 691]$$

and reserve capacities (pcu/h)

$$[RC^{(SC)}] = [159\ 489\ 153\ 0]$$

It is worth noting that, under this traffic condition, the reserve capacity is zero for the leg (in the case under examination, leg n° 4) that reaches the saturation state first.

2.6.1.2 Total Capacity Calculation

To calculate total capacity, we can randomly choose the first-attempt, entering flow vector.

For the case under examination, we assume

$$[Q_{ei}^{(1)}] = [100\ 220\ 300\ 300]$$

Then, we determine the circulating volume in front of entry 1

$$Q_{c1}^{(1)} = (P_{42} + P_{43}) \cdot Q_{e4}^{(1)} + P_{32} \cdot Q_{e3}^{(1)} = (0.74 + 0.07) \cdot 300 + 0.15 \cdot 300 = 288$$

and, by the capacity formula, a new entering flow value at leg 1 (in pcu/h)

$$Q_{e1}^{(2)} = 1218 - 0.74 \cdot Q_{c1}^{(1)} = 1218 - 0.74 \cdot 288 = 1005\ \text{pcu/h}$$

We proceed similarly for entries 2, 3, and 4. We use the "most updated" entering traffic flows coming from the various legs to calculate the conflict flows:

$$Q_{c2}^{(1)} = (P_{13} + P_{14}) \cdot Q_{e1}^{(2)} + P_{43} \cdot Q_{e4}^{(1)} = (0.75 + 0.10) \cdot 1005 + 0.07 \cdot 300 = 875\,\text{pcu/h}$$

$$Q_{e2}^{(2)} = 1218 - 0.74 \cdot Q_{c2}^{(1)} = 1218 - 0.74 \cdot 875 = 570\,\text{pcu/h}$$

$$Q_{c3}^{(1)} = (P_{24} + P_{21}) \cdot Q_{e2}^{(2)} + P_{14} \cdot Q_{e1}^{(2)} = (0.57 + 0.19) \cdot 570 + 0.10 \cdot 1005 = 534\,\text{pcu/h}$$

$$Q_{e3}^{(2)} = 1218 - 0.74 \cdot Q_{c3}^{(1)} = 1218 - 0.74 \cdot 534 = 823\,\text{pcu/h}$$

$$Q_{c4}^{(1)} = (P_{31} + P_{32}) \cdot Q_{e3}^{(2)} + P_{21} \cdot Q_{e2}^{(2)} = (0.63 + 0.15) \cdot 823 + 0.19 \cdot 570 = 750\,\text{pcu/h}$$

$$Q_{e4}^{(2)} = 1218 - 0.74 \cdot Q_{c4}^{(1)} = 1218 - 0.74 \cdot 750 = 663\,\text{pcu/h}$$

Thus, we obtain a second entering flow vector (pcu/h)

$$[Q_{ei}^{(2)}] = [1005\ 570\ 823\ 663]$$

We proceed similarly, and we obtain

$$[Q_{ei}^{(3)}] = [729\ \ 725\ \ 756\ \ 680]$$

$$[Q_{ei}^{(4)}] = [727\ \ 726\ \ 756\ \ 680]$$

$$[Q_{ei}^{(5)}] = [727\ \ 726\ \ 756\ \ 680]$$

The iterative process converges very rapidly. The entering flow vectors determined for steps (4) and (5) are equal, and they represent the number of vehicles that each leg is able to serve when all the legs are saturated.

The sum of all these flows provides the roundabout total capacity TC, which is 2888 pcu/h.

References

1. *Highway Capacity Manual HCM 2000*, Special Report n. 209, T.R.B., Washington DC, 2000.
2. Brilon, W., Wu, N., Bondzio, L., *Unsignalized Intersections in Germany – A State of the Art 1997*, Proceedings of the Third International Symposium on Intersections Without Traffic Signals, pp. 61–70, Portland, Oregon, USA, July 1997.
3. Brilon, W. (Editor), *Intersections Without Traffic Signals*, Vol. II, Springer-Verlag, Berlin, 1991.
4. Kimber, R. M., *The Traffic Capacity of Roundabouts*, TRRL Laboratory, Report 942, Crowthorne, Berkshire, 1980.

5. Louah, G., *Panorama critique des modeles français de capacité des carrefours giratoires*, Actes du séminaire international "Giratoires 92," Nantes, France, 14–16 Octobre 1992.
6. Guichet, B., *Roundabouts in France: Development, Safety, Design and Capacity*, Proceedings of the Third International Symposium on Intersections Without Traffic Signals, pp. 100–105, Portland, Oregon, USA, July 1997.
7. Guichet, B., *Roundabouts in France and New Use*, TRB Transportation Research, National Roundabout Conference Proceedings 2005, Vail, Colorado, USA, 22–25 May 2005.
8. CERTU, *Carrefours urbains Guide*, Lyon, France, 1999.
9. *Merkblatt für die Anlage von Kreisverkehren*, FGSV Verlag GmbH, Köln, Germany, 2006.
10. Marlow, M., Maycock, G., *The Effect of Zebra Crossings on Junction Entry Capacities*, Report SR 742, Transport and Road Research Laboratory (TRRL), Crowthorne, Berkshire, England, 1982.
11. Brilon, W., Stuwe, B., Drews, O., *Sicherheit und Leistungsfähigkeit von Kreisverkehrsplätzen*, Institute for Traffic Engineering, Ruhr Universität, Bochum, Deutschland, 1993.

Chapter 3
Waiting Phenomena at Steady State and Non-steady State Conditions

To calculate waiting times, queue lengths, and the number of vehicles in the system,[1] it is possible to consider roundabout entry lanes as channels of an ordinary system examined by means of the mathematical queuing theory (Fig. 3.1).

Therefore, for the elements shown in Fig. 3.1 and for the specific case being considered here, we have that[2]:

- the arrival and departure processes, which are made up of a events that are generally represented by random instants (flows of events), are characterized on the basis of their respective traffic count laws [1];
- the service point is located on the entry yielding lines.

The service method (service mechanism) consists of waiting for a time gap in the traffic stream flowing on the circulatory roadway that is sufficient to insert one or more vehicles in sequence into the roundabout (gap-acceptance [2]). On the basis of this service mechanism, the service time proprieties T_s are established in probabilistic terms (probabilistic distribution).

- the waiting organization (queue discipline) requires that the users be served on the basis of their arrival time in the system (sometimes called "first come, first served" (FCFS), but more frequently referred to as "first in, first out" (FIFO);
- the number of queuing users can be high (unlimited longitudinal capacity of the channel). Once in the queue, a vehicle cannot leave it without entering and exiting from the roundabout.

In the remaining part of this chapter, first, we will review some results of stochastic and deterministic theory about waiting phenomena.

Then, we will present a general heuristic criterion that allows the evaluation of the system state variables at any operational conditions of the entries (i.e., undersaturation, saturation, and oversaturation).

---

[1] We recall that the definitions of these state variables were given in Sect. 1.3.

[2] Examining the elements shown in Fig. 3.1, the queuing theory terms are reported in brackets.

R. Mauro, *Calculation of Roundabouts*, DOI 10.1007/978-3-642-04551-6_3,

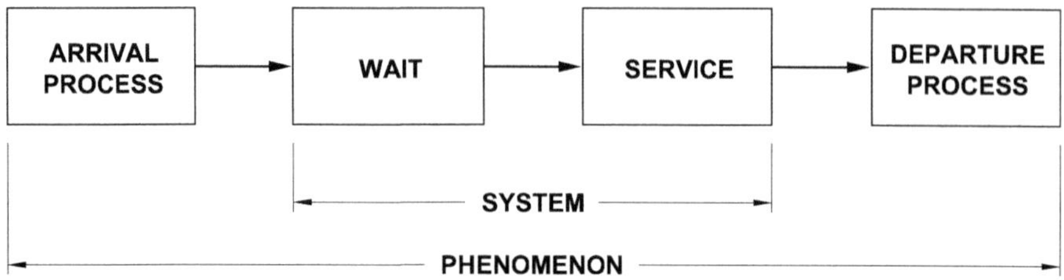

Fig. 3.1 Schematic diagram of a queuing system

The above-mentioned method uses transition relationships between results obtained with the probabilistic and deterministic approaches.

Finally, in the next chapter, the criteria and the methods are applied to the study of roundabouts that have time-dependent traffic demands; in particular, we will consider roundabouts with traffic peaks that occur between two steady-state periods of flow.

3.1 Some Results of Probabilistic Analysis of Queues

From the probabilistic theory of queuing, we will now review some definitions and fundamental relationships for the applications.

We define traffic intensity or degree of saturation as the ratio ρ of traffic demand Q_e at an entry to capacity C of the same entry[3]

$$\rho = \frac{Q_e}{C} \tag{3.1}$$

If $E[T_S]$ is the mean of service time T_S, given the meaning of C and T_S, we can obtain

$$E[T_S] = \frac{1}{C} \tag{3.2}$$

Recalling Eq. (1.17) of Chap. 1, by which we defined reserve capacity ($RC = C - Q_e$), from Eqs. (3.1) and (3.2), we have

$$\rho = 1 - \frac{RC}{C} \tag{3.3}$$

Therefore, the entry condition can be characterized in terms of RC or of ρ as shown in Table 3.1.

[3] For simplicity notation, in this chapter we omit the subscripts to the flows and capacities relative to the generic entry "i".

Table 3.1 Reserve capacity RC, degree of saturation ρ and entry states conditions

Undersaturated entry	Saturated entry	Oversaturated entry
$RC > 0$ $\rho < 1$	$RC = 0$ $\rho = 1$	$RC < 0$ $\rho > 1$

If we indicate the average time spent in the system with $E[w_S]$, using Eq. (1.20) of Chap. 1 ($w_S = w_c + T_S$) and Eq. (3.2), above, we have the following equation for the propriety of the mean as a linear operator:

$$E[w_S] = E[w_c] + E[T_S] = E[w_c] + \frac{1}{C} \tag{3.4}$$

In addition, to calculate the average number $E[L_S]$ of vehicles waiting in the system, it is necessary to know the relationship that exists between $E[L_S]$, Q_e, and $E[w_S]$ under steady-state conditions (first formula by Little):

$$E[L_S] = Q_e \cdot E[w_S] \tag{3.5}$$

Equation (3.5) can briefly be accounted for as follows. During the waiting time w_S of a generic vehicle, the number of vehicles that enter the system is equal to:

$$L_S = Q_e \cdot w_S \tag{3.6}$$

Under conditions of statistical equilibrium, Q_e results are time-independent, so we can average the terms of Eq. (3.6) to obtain Eq. (3.5).

A relationship that is homologous to Eq. (3.6) can be written for the queue length L_c, obtaining $L_c = Q_e \cdot w_c$.

3.1.1 Some Remarkable Results

We will now present some results of the queuing theory relative to unsignalized intersections at statistical equilibrium (steady-state conditions of the system).

First, we must characterize the demand flow Q_e probabilistically and specify the service mechanism on which the determination of service time T_S depends.

In this connection, it is possible to use for roundabouts – as for any type of unsignalized intersections – different methods for evaluating T_S.

For their detailed descriptions, see [2], where each method for calculating T_S is presented, along with the formulas for the averages $E[w_S]$ and $E[L_S]$. In current practice, however, the most widespread service mechanism is the one described below.

Consider A and B to be the first two vehicles in the system. Assume that vehicle A uses an interval Δt equal to or slightly greater than the critical gap T_c for its maneuver. Vehicle B's time T_S occurs between the moment that it arrives at the yielding line and the moment it enters the circulatory roadway. Thus, for B, a time elapses that is equal to the sum of: the interval $\Delta t - T_c$ + the possible successive gaps between the circulating vehicles rejected by B + the critical gap T_c. This sum is the service time T_s of vehicle B.

If T_s is determined in this way, it is not possible to model, for more than one vehicle, a service that is provided with continuity inside the same time interval and in sequence, i.e. without interruptions of the entering process into the circulatory roadway.

With the service procedures just recalled, the well-known P-K relationships (from the authors Pollaczeck-Khinchine) [3] relative to queuing theory can be easily used. According to these relationships:

$$E[w_S] = E[T_s] + \frac{Q_e \cdot (\overline{E[T_S]}^2 + VAR[T_S])}{2 \cdot (1 - Q_e \cdot E[T_s])} \tag{3.7}$$

where $E[T_S]$ and $VAR[T_S]$ are the mean and the variance of service time, respectively. If we assume a T_s distribution according to the Erlang function with a parameter k, these two variables can be obtained from the following equations:

$$E[T_s] = T_c + \frac{e^{kQ_cT_c} - \sum_{i=0}^{k} \frac{(kQ_cT_c)^i}{i!}}{Q_c \sum_{i=0}^{k} \frac{(kQ_cT_c)^i}{i!}} \tag{3.8}$$

$$VAR[T_s] = \frac{(k+1) \cdot \left[e^{kQ_cT_c} - \sum_{i=0}^{k+1} \frac{(kQ_cT_c)^i}{i!} \right]}{kQ_c^2 \sum_{i=0}^{k-1} \frac{(kQ_cT_c)^i}{i!}} + (E[T_s] - T_c)^2 \tag{3.9}$$

The parameter k of the Erlang function must be calculated on the basis of the traffic flow rate Q_c, which can be approximated according to the indications shown in Table 3.2.

Remember that Q_c for a roundabout is the circulating flow in front of the entry considered (See Fig. 1.2 in Chap. 1).

Table 3.2 Parameter k value of the Erlang probability law versus of circulating flow Q_c

Flow rate Q_c [veh/h]	K value
0–500	1
501–1000	2
1001–1500	3

The relationship shown in Table 3.2 was determined experimentally. If Q_c (expressed as hourly volume) is known, the relationship can be used to determine k as an integer value for intervals of Q_c.

Equations (3.7), (3.8), and (3.9) are valid for Poissonian arrivals into the secondary stream Q_e and service time T_s distributed in any way (depending on T_c and on the characteristics of the main stream Q_c, represented by the value selected for k).

Evaluating $E[w_s]$ by means of Eq. (3.7), the average number $E[L_s]$ of vehicles in the system is given by Eq. (3.5) on the basis of which:

$$E[L_s] = Q_e \cdot E[w_s] = Q_e \cdot E[T_s] + \frac{Q_e^2 \cdot (\overline{E[T_s]}^2 + VAR[T_s])}{2 \cdot (1 - Q_e \cdot E[T_s])} \tag{3.10}$$

The non-approximated determination of the probability distribution L_s is not easy to represent in closed form.

However, it is worth noting that doubling $E[L_s]$ generally provides a high enough percentile ($L_{s,90}$; $L_{s,95}$) estimation of L_s for the applications.

Finally, we recall that the results and the considerations outlined so far to calculate $E[w_s]$ are always relative to the simple case of only two conflicting flows (Q_e and Q_c) or to schemes that can be assimilated to it (See Fig. 1.2 in Chap. 1).

A simple, concise deduction of Eqs. (3.7) and (3.10) is reported in [3].

Equations (3.7) and (3.10) are now specified for two cases that are important for practical applications.

First of all, we simplify the notations in Eq. (3.7) as:

$$T_s = s; E[T_s] = \overline{s}; VAR\,[T_s] = V[s]$$

Therefore, we rewrite Eq. (3.7) as:

$$E[w_s] = \overline{s} + \frac{Q_e \cdot (\overline{s}^2 + V[s])}{2 \cdot (1 - Q_e \cdot \overline{s})} \tag{3.11}$$

which, with Eqs. (3.1) and (3.2), becomes:

$$E[w_s] = \overline{s} + \frac{\rho \cdot \left(\overline{s} + \dfrac{V[s]}{\overline{s}}\right)}{2 \cdot (1 - \rho)} \tag{3.12}$$

From Eq. (3.10), we have for the average number $E[L_s]$ of vehicles in the system:

$$E[L_s] = Q_e \cdot \overline{s} + \frac{Q_e^2 \cdot (\overline{s}^2 + V[s])}{2 \cdot (1 - Q_e \cdot \overline{s})} \tag{3.13}$$

which, again for Eqs. (3.1) and (3.2), is:

$$E[L_S] = \rho + \frac{\rho^2 \cdot \left(1 + \frac{V[s]}{\bar{s}^2}\right)}{2 \cdot (1 - \rho)} \tag{3.14}$$

As we will see for some cases, with Eqs. (3.12) and (3.14) we have a straightforward characterization of the expressions of $E[w_S]$ and $E[L_S]$.

3.1.1.1 Poissonian Arrivals and Exponential Service Times

Under this assumption, since the probability density function $f_{TS}(t)$ of the service time T_S, is exponential, we have the following relationship between mean and variance:

$$V[s] = \bar{s}^2 \tag{3.15}$$

Then, Eq. (3.12) becomes:

$$E[w_S] = \bar{s} + \frac{\rho\left(\bar{s} + \frac{\bar{s}^2}{\bar{s}}\right)}{2(1 - \rho)} = \frac{2\bar{s} - 2\rho\bar{s} + 2\rho\bar{s}}{2(1 - \rho)} = \frac{\bar{s}}{(1 - \rho)} \tag{3.16}$$

Finally, using Eq. (3.2), we have:

$$E[w_S] = \frac{1}{C \cdot (1 - \rho)} \tag{3.17}$$

From Eq. (3.14), we have for $E[L_S]$:

$$E[L_S] = \rho + \frac{\rho^2 \cdot \left(1 + \frac{\bar{s}^2}{\bar{s}^2}\right)}{2(1 - \rho)} = \frac{\rho - \rho^2 + \rho^2}{1 - \rho} = \frac{\rho}{1 - \rho} \tag{3.18}$$

3.1.1.2 Poissonian Arrivals and Deterministic Service Times

This occurs when the service is regular, i.e., each waiting vehicle spends the same time at the head of the queue to enter the circulatory roadway.

This means that

$$V[s] = 0 \tag{3.19}$$

Then, taking into account Eq. (3.2), Eq. (3.12) becomes:

$$E[w_S] = \bar{s} + \frac{\rho \cdot \bar{s}}{2 \cdot (1 - \rho)} = \frac{1}{C} \cdot \left[\frac{2 - \rho}{2 \cdot (1 - \rho)}\right] \tag{3.20}$$

Table 3.3 Average values of state variables according to statistical equilibrium

	Queue		System	
	$E[w_c]$	$E[L_c]$	$E[w_s]$	$E[L_s]$
Poissonian arrivals; exponential service times	$\frac{\rho}{C \cdot (1-\rho)}$	$\frac{\rho^2}{(1-\rho)}$	$\frac{1}{C \cdot (1-\rho)}$	$\frac{\rho}{(1-\rho)}$
Poissonian arrivals; constant service times	$\frac{\rho}{2 \cdot C \cdot (1-\rho)}$	$\frac{\rho^2}{2 \cdot (1-\rho)}$	$\frac{2-\rho}{2 \cdot C \cdot (1-\rho)}$	$\frac{2\rho-\rho^2}{2 \cdot (1-\rho)}$

From Eq. (3.14), we can write for $E[L_s]$:

$$E[L_s] = \rho + \frac{\rho^2}{2(1-\rho)} = \frac{2\rho - 2\rho^2 + \rho^2}{2(1-\rho)} = \frac{2\rho - \rho^2}{2(1-\rho)} \tag{3.21}$$

Starting once again with Eqs. (3.7) and (3.10), expressions for the averages $E[w_c]$ and $E[L_c]$ can be easily derived for the waiting time in the queue and queue length.

The expressions of the average values so far recalled are shown in Table 3.3.

3.1.2 Concluding Remarks

To conclude this section, Fig. 3.2 shows the graphs of Eqs. (3.17) and (3.18).

We can note that the behavior of the average time spent in the system $E[w_s]$ and of the average number of users in the system $E[L_s]$ are similar (as is predictable on the basis of Eq. (3.5)). They are both characterized by a knee, beyond which, the

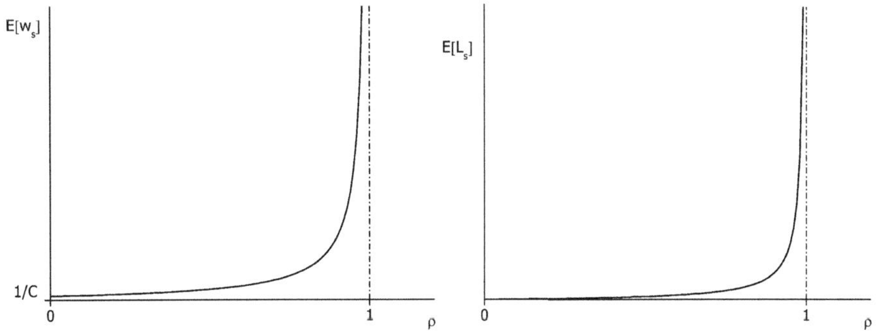

Fig. 3.2 Averages $E[w_s]$ and $E[L_s]$ versus degree of saturation ρ (Poissonian arrivals; exponential service times; one-queue system)

gradients grow rapidly when traffic intensity ρ increases, and a vertical asymptote appears when $\rho = 1$

$$\lim_{\rho \to 1} E[w_S] = \infty \tag{3.22}$$

$$\lim_{\rho \to 1} E[L_s] = \infty \tag{3.23}$$

We have similar behaviors for the queue, for $E[w_C]$, and for $E[L_C]$.

We can easily note that, if $E[w_S]$ and $E[L_S]$ are expressed as a function of reserve capacity, we can use Eqs. (3.17) and (3.18) to obtain the curves shown in Fig. 3.3, where the asymptote coincides with the ordinate axis.

As we will soon see with an example, this behavior of $E[w_S]$ and $E[L_S]$ derives directly from the two basic conditions of stationariness, i.e., from the assumption that the entering traffic demand remains constant indefinitely.

In addition, this asymptotic behavior of $E[w_S]$ and $E[L_S]$ is not only characteristic of queuing systems with Poissonian arrivals and exponential service times.

In fact, all the queuing systems that may be modeled by Eqs. (3.7) and (3.10) show a similar behavior for $E[w_S]$ and $E[L_S]$, as can be easily seen from Eqs. (3.12) and (3.14) for ρ which tends to be unity.

Consider an entry that is systematically saturated, i.e., $Q_e(t) = C(t)$ and $Q_e(t+\Delta t) = Q_e(t)$ for any Δt.

For example, Q_e is equal to 1440 veh/h = 0.40 veh/s, i.e., on average, 0.4 arriving vehicles are recorded for each second.

$C = Q_e$ indicates that there is, on average, 0.4 vehicle served (in the case vehicles entering the circulatory roadway of a roundabout) for each second used for the service.

For $T = 10000$ s, about 4000 arrivals are recorded. Due to the random variations of the system, the arrivals of 4000 vehicles will not be distributed uniformly over T, which means that sometimes there will be a queue, and sometimes there will not be a queue.

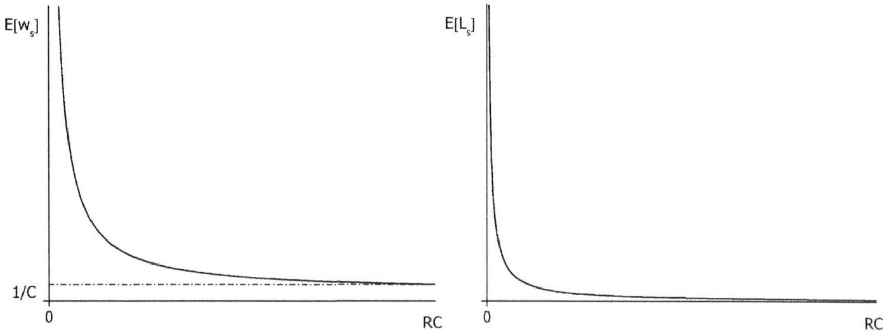

Fig. 3.3 Averages $E[w_S]$ and $E[L_S]$ versus Reserve Capacity RC (Poissonian arrivals; exponential service times; one-queue system)

For the same reasons, the time gaps used for the service will have different values and will not be localized uniformly in T = 10000 s of system observation. In addition, their overall duration will be shorter than T.

For example, if we assume that the overall duration is 9000 s, the number of arrivals into the roundabout is therefore of $9000 \cdot C = 9000 \cdot 0.4 = 3600$ vehicles.

In conclusion, at the end of the T during which 4000 vehicles approach the intersection along the leg considered, and 3600 vehicles enter the roundabout, there would be 400 vehicles waiting in the queue for entry, which is an estimation of $E[L_S]$.

It is now straightforward to understand that, when all the other conditions are equal, if T increases indefinitely, even $E[L_S]$ tends to increase indefinitely (Eq. (3.23)). This is consistent with the presence of the asymptote that occurs when $\rho = 1$, as predicted by Eq. (3.17) (See Fig. 3.2).

From Eq. (3.5), the presence of the asymptote when $\rho = 1$ is also justified for $E[w_S]$.

3.2 Deterministic Analysis of Queues

The analyses are based on the cumulative time value t of arrivals (at legs) A(t) and departures D(t).

A departure from the queue occurs when a vehicle waiting in second position moves to the head of the queue, close to the yielding line.

When there is no queue, the term departure in this context also refers to the travel of a vehicle from the head of the traffic stream to the waiting position to enter the intersection.

In conclusion (See Fig. 3.4), a departure is recorded when a vehicle crosses the imaginary line (a–a) that marks the upstream limit to the service position.

Since traffic counts are discrete, A(t) and D(t) are step functions. Instead of these, the respective continuous approximations are generally used (fluid approximations, Fig. 3.5).

The continuous approximation derives from the continuous description (or continuous at intervals) of traffic demand $Q_e(t)$ and entry capacity C(t).

From the definition of volume, the arrival rate is therefore equal to

$$\frac{dA(t)}{dt} = Q_e(t) \quad \forall t \tag{3.24}$$

With the deterministic approach, $Q_e(t)$ and C(t) are not subject to random variations.

Under this assumption:

- the determinations of Q_e and C coincide with the mean values

$$Q_e(t) = E[Q_e(t)]; \quad C(t) = E[C(t)] \quad \forall t \tag{3.25}$$

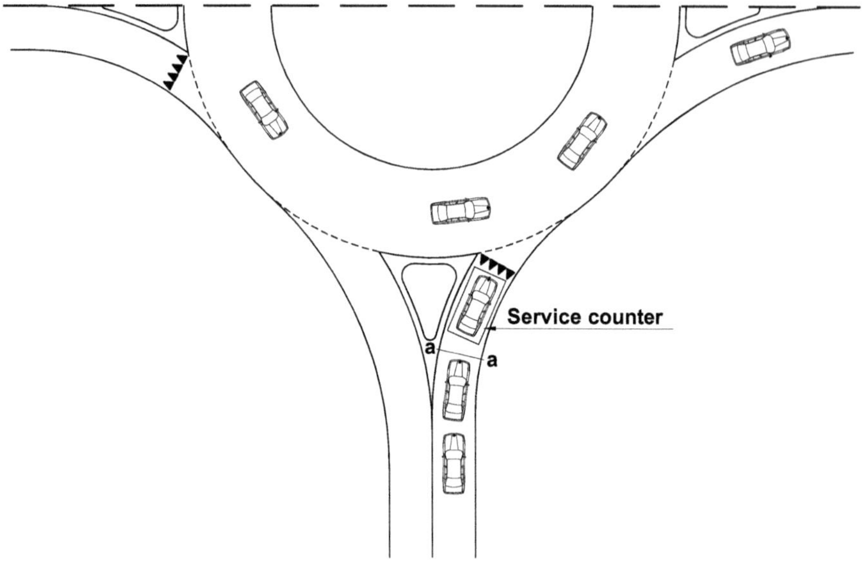

Fig. 3.4 Priority system and service counter

– for the cumulative rate of departures it is straightforward to put:

$$\text{a)} \quad \frac{dD(t)}{dt} = Q_D(t) = Q_e(t) \qquad t \in (t_i; t_j) \tag{3.26}$$

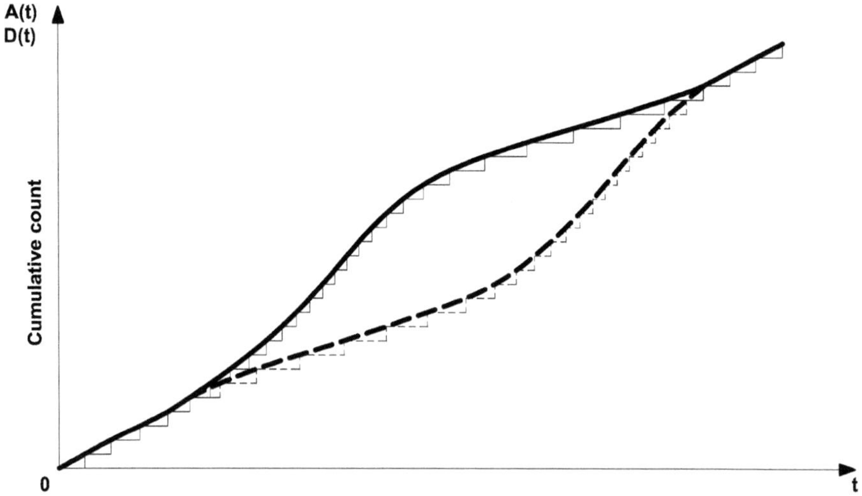

Fig. 3.5 Fluid approximation of the cumulative arrival and departure counts

when, during an entire generic interval (t_i; t_j), the entry is undersaturated (e.g. $Q_e(t) < C(t)$) and at the initial instant there is no queue at the entry (e.g. $L_c(t_i) = 0$)

$$\text{b)} \quad \frac{dD(t)}{dt} = Q_D(t) = C(t) \qquad t \in (t_i; t_j) \tag{3.27}$$

if during the entire generic interval (t_i; t_j), the entry is saturated or oversaturated (e.g. $Q_e(t) \geq C(t)$) or if at the initial instant the queue L_0 is present at entry (e.g. $L_c(t_i) = L_0$). If $L_c(t_i) = L_0$ and $C(t) > Q_e(t)$, $t_j = f(L_0; C(t))$.

Once the behaviors of $Q_e(t)$ and $C(t)$ are known, $A(t)$ and $D(t)$ are obtained by integration from Eqs. (3.24), (3.26), and (3.27).

Starting with the curves $A(t)$ and $D(t)$ (See Fig. 3.6), the main indicators of the waiting phenomenon can be calculated directly.

The length of the queue $L_c(t)$ at a generic instant t is obtained as

$$L_c(t) = A(t) - D(t) \tag{3.28}$$

Equation (3.28) expresses the preservation law of vehicles at legs, on the basis of which the arriving vehicles can only leave the queue or stay in it ($A(t) = D(t) + L_c(t)$).

As the queue method for entering the intersection is of the FIFO type (See Sect. 1.3), the horizontal distance d between the curves $A(t)$ and $D(t)$ is the waiting time in the queue w_c for the vehicle that arrived at time t:

$$d = w_c(t) \tag{3.29}$$

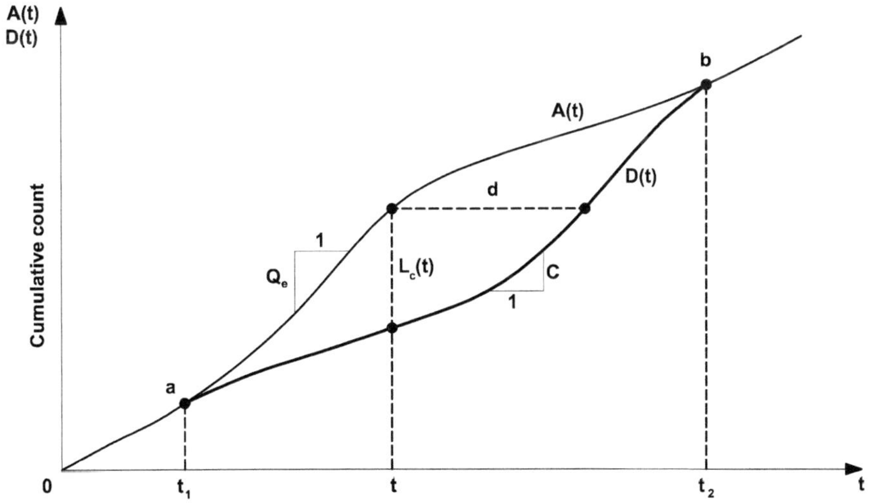

Fig. 3.6 Cumulative arrival count A(t) and cumulative departure count D(t)

It follows that the area S(aba) included between A(t) and D(t) is the total waiting time in the queue W_c, i.e., regarding all the users who have entered the queue.

The calculation of W_c allows the evaluation of the mean waiting time $\overline{w}_c$ as a ratio between W_c and a suitable number of users represented by vehicles relative to the pre-fixed time $T_q = t_i - t_f$:

$$\bar{w}_c = \frac{W_c}{\int_{t_i}^{t_f} Q_e(t)dt} = \frac{\int_{t_i}^{t_f} [A(t) - D(t)]dt}{\int_{t_i}^{t_f} Q_e(t)dt} \tag{3.30}$$

The use of the functions A(t) and D(t) also allows, as we will later see with a worked example, the determination of the time T_d inside which the effects of the peak traffic can be observed, when this same peak is ended.

In the above-mentioned example, as it generally occurs in the analysis of the performance indices of intersections, the demand Q_e and capacity C are modeled with step functions.

With Eqs. (3.24), (3.26), and (3.27), broken lines are obtained by which the calculation (and implementation of the results which it gives) of the cumulative arrival and departure values are very straightforward due to the use of elementary geometrical considerations.

So far, all that has been said about departures from the queue can be repeated for departures from the system (See Fig. 1.5, Chap. 1). In fact, the waiting time in the queue w_c, as we have already seen, is equal to the horizontal distance between the cumulative counts A(t) and D(t) (See Fig. 3.6); in addition, the relationship between the time spent in the system w_s, service time T_s, and queuing time w_c is expressed by Eq. (1.20) from Chap. 1.

$$w_s = w_c + T_s = w_c + \frac{1}{C} \tag{3.31}$$

The cumulative count $D_s(t)$ of the departures from the system is therefore obtained instant by instant by moving the value of D(t) horizontally by a value equal to $T_s = 1/C$.

3.2.1 A Worked Example

Consider an entry with a variable traffic demand. Capacity during the observation period has a constant behavior.

To simplify the notation, the subscript for traffic demand, i.e., "e," is omitted.

Initial Data

The arrival flow rate at the legs is $Q_1 = 1600$ veh/h from 6:45 to 7:00 AM. In addition, it results that $Q_1 < C = 2000$ veh/h.

During the period from 7:00 to 7:15 AM, the arrival rate increases to $Q_2 = 2400$ veh/h, so $Q_2 > C$.

From 7:15 to 7:30 AM, demand becomes $Q_3 = 2200$ veh/h, and $Q_3 > C$.

Finally, between 7:30 and 7:45 AM the arrival rate decreases to $Q_4 = 1200$ veh/h, so $Q_4 < C$.

The volumes Q_1, Q_2, Q_3, and Q_4 and capacity C (rate of flowing vehicles) shown in Fig. 3.7a cause (relationships (3.24), (3.26), and (3.27)), the curves of the cumulative arrival values and of the cumulative departure values of the queue as functions of time. The curves are represented, respectively, by the line indicated as "arrivals" and by the "departures" line shown in Fig. 3.7b on the basis of the relationships:

$$\text{Cumulative value of demand (arrivals)} = Q \cdot t \tag{3.32}$$

$$\text{Cumulative count of departures from the queue} = C \cdot t \tag{3.33}$$

Number of Arriving Vehicles

The number of arriving vehicles between 6:45 and 7:00 AM is:

$$A_1 = Q_1 \cdot t_1 = 1600\,\text{veh/h} \cdot (0.25\,\text{h}) = 400\,\text{veh}$$

This number is equal to the area B_1 of the bar diagram shown in Fig. 3.7a. Similarly, for each of the remaining analysis periods, the cumulative counts A_2, A_3, and A_4 are determined for a total of 1850 vehicles in the 45 min considered.

Time of the Beginning of the Congestion

Congestion begins exactly when demand exceeds capacity (i.e., when the entry is oversaturated), that is, when $Q_2 > C$. This event is recorded at 7:00 AM.

Congestion Duration ($T_q = t_f - t_i$)

During the periods t_2 and t_3, the arriving flows exceed capacity C, while the queue disappears in the interval t_4 (See Fig. 3.7b) and remains for the duration T_d (part of t_4):

$$T_q = t_2 + t_3 + T_d \tag{3.34}$$

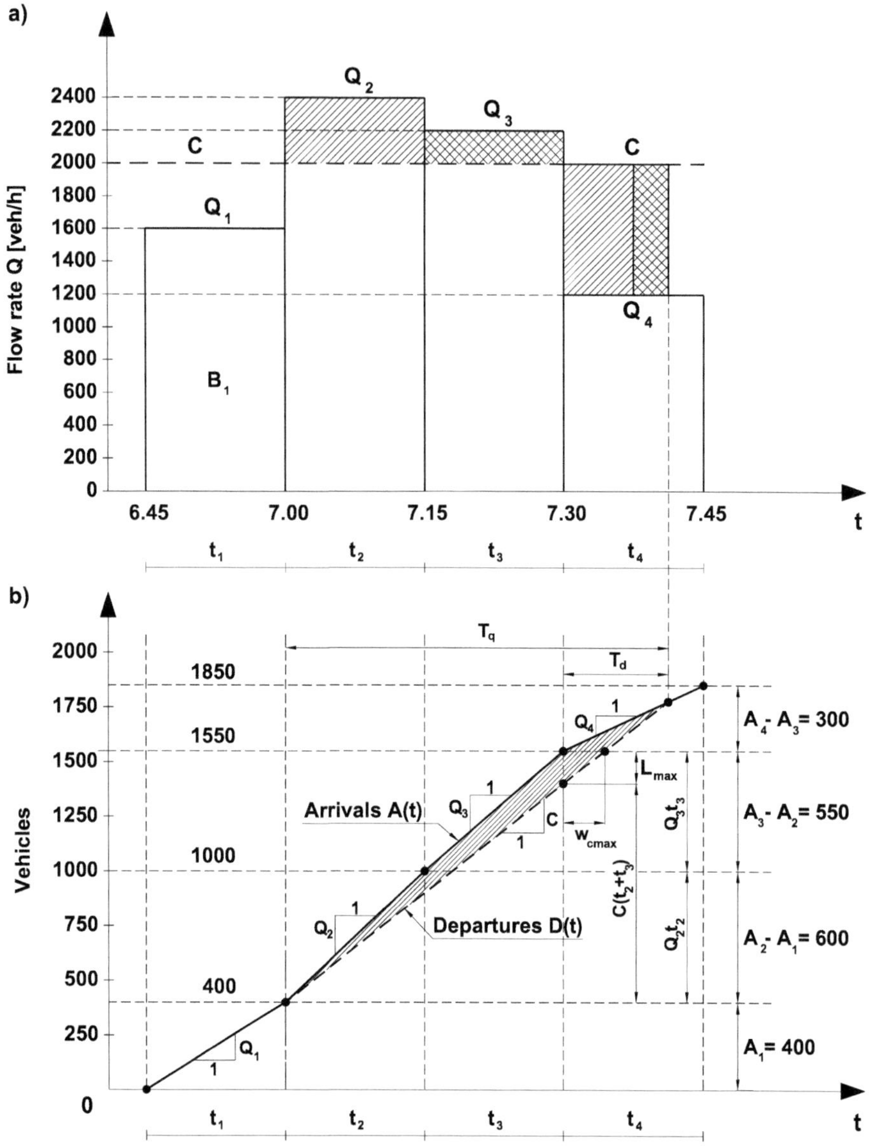

Fig. 3.7 Demand flow rate, cumulative arrival count and cumulative departure count at leg

In addition, in the unknown time T_q, the following vehicles must be cleared: $(Q_2–C)\cdot t_2$; $(Q_3–C)\cdot t_3$; and $Q_4\cdot T_d$; that is in all:

$$(Q_2 - C) \cdot t_2 + (Q_3 - C) \cdot t_3 + Q_4 \cdot T_d$$

As in T_d, the queue L_{max}, present at the end of t_3, is cleared

$$(Q_2 - C) \cdot t_2 + (Q_3 - C) \cdot t_3 + (C - Q_4) \cdot T_d = 0$$

and, therefore,

$$T_d = \frac{(Q_2 - C) \cdot t_2 + (Q_3 - C) \cdot t_3}{C - Q_4}$$

which, substituted into Eq. (3.34), gives

$$T_q = t_2 + t_3 + \frac{(Q_2 - C) \cdot t_2 + (Q_3 - C) \cdot t_3}{C - Q_4} =$$

$$= 0.25 + 0.25 + \frac{(2400 - 2000) \cdot 0.25 + (2200 - 2000) \cdot 0.25}{2000 - 1200} = 0.69 \text{ h} = 41.4 \text{ min}$$

In other words, the queue that began at 7:00 o'clock AM disappears at 7:41:24 AM.

Maximum Number of Queuing Vehicles L_{max}

The maximum number of queuing vehicles is recorded at the end of period t_3 and is equal to the cumulative value of demand rate that was not cleared during the period $t_2 + t_3$.

$$L_{max} = (Q_2 - C) \cdot t_2 + (Q_3 - C) \cdot t_3 = $$
$$= (2400 - 2000) \cdot 0.25 + (2200 - 2000) \cdot 0.25 = 150 \text{ veh}$$

Maximum Duration of the Individual Delay w_{cmax}

This delay refers to the vehicle that arrives at the end of the period t_3, that is at 7:15 AM, and is equal to (See Fig. 3.7b)

$$w_{cmax} = \frac{Q_2 \cdot t_2 + Q_3 \cdot t_3}{C} - t_2 - t_3 = \frac{Q_2 \cdot t_2 + Q_3 \cdot t_3 - C \cdot t_2 - C \cdot t_3}{C} =$$

$$= \frac{(Q_2 - C) \cdot t_2 + (Q_3 - C) \cdot t_3}{C} = \frac{(2400 - 2000) \cdot 0.25 + (2200 - 2000) \cdot 0.25}{2000} =$$
$$= 0.075 \text{ h} = 4.5 \text{ min}$$

Total Duration of the Delay W_c

W_c is represented by the area S of the cumulative diagram (Q_e; C).

$$W_c = S = \frac{(Q_2 t_2 - C t_2) \cdot t_2}{2} + \frac{(Q_2 t_2 - C t_2) \cdot t_3}{2} + \frac{[(Q_2 - C) \cdot t_2 + (Q_3 - C) \cdot t_3]}{2} \cdot t_3 +$$
$$+ \frac{[(Q_2 - C) \cdot t_2 + (Q_3 - C) \cdot t_3]\left[\frac{(Q_2 - C) \cdot t_2 + (Q_3 - C) \cdot t_3}{C - Q_4}\right]}{2} =$$
$$= \frac{(Q_2 t_2 - C t_2) \cdot (t_2 + t_3)}{2} + \frac{(Q_2 - C) \cdot t_2 + (Q_3 - C) \cdot t_3}{2}\left[t_3 + \frac{(Q_2 - C) \cdot t_2 + (Q_3 - C) \cdot t_3}{C - Q_4}\right] =$$

$$= \frac{[2400 \cdot 0.25 - 2000 \cdot 0.25] \cdot (0.25 + 0.25)}{2} + \frac{(2400 - 2000) \cdot 0.25 + (2200 - 2000) \cdot 0.25}{2} \cdot$$
$$\cdot \left[0.25 + \frac{(2400 - 2000) \cdot 0.25 + (2200 - 2000) \cdot 0.25}{2000 - 1200}\right] = 57.81 \text{ (h veh)}$$

Number of Delayed Vehicles N_R

$$N_R = C \cdot T_q = 2000 \cdot 0.69 = 1380\,\text{veh}$$

Mean Delay for Each Vehicle Affected by the Disturbance $\overline{w}_c$

$$\overline{w}_c = \frac{W_c}{N_R} = \frac{57.81}{1380} = 0.041\,\text{h} \cong 2.5\,\text{min}$$

Average Length of the Queue $\overline{L}_c$

$$\overline{L}_c = \frac{W_c}{T_q} = \frac{57.81}{0.69} \cong 84\ \text{veh}$$

3.2.2 Some Remarkable Results

The deterministic approach is used to analyze two important cases for the applications of practical interest.

3.2.2.1 First Remarkable Case and a Worked Example

Assume (See Fig. 3.8) that at the initial instant t_0 of system observation, $L_c(t_0) = L_{c0}$ vehicles are present in the queue.

Traffic demand $Q_e(t)$ and capacity $C(t)$ starting with t_0 are systematically assumed to be constant ($Q_e(t) = Q_e$ and $C(t) = C$), with $Q_e > C$ (oversaturated entry).

At the instant t, the cumulative arrival and departure counts are, on the basis of Eqs. (3.24) and (3.27),

$$A(t) = A(t_0) + Q_e \cdot T \tag{3.35}$$

$$D(t) = D(t_0) + C \cdot T \tag{3.36}$$

Since $A(t_0) - D(t_0) = L_{c0}$, we can use Eqs. (3.28), (3.35), and (3.36) to produce

$$L_c(t) = L_{c0} + (Q_e - C) \cdot T \tag{3.37}$$

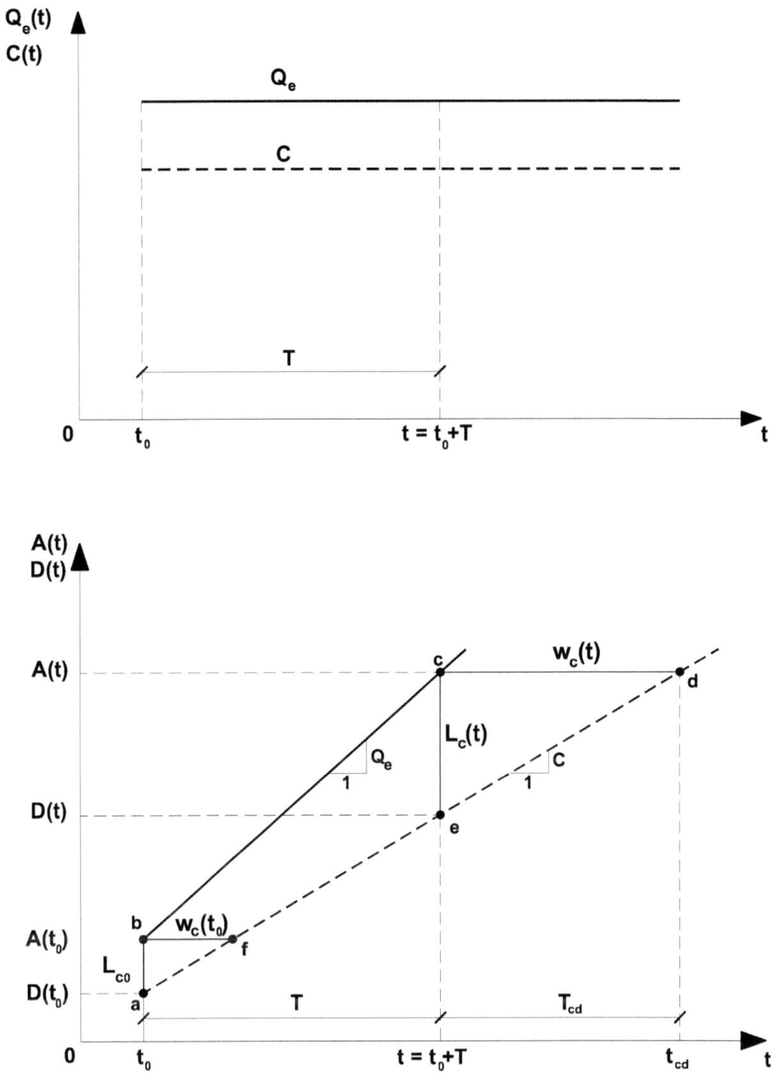

Fig. 3.8 Traffic demand, capacity, and cumulative arrival and departure values regarding the first remarkable case

Recalling Eq. (3.1), Eq. (3.37) becomes

$$L_c(t) = L_{c0} + (\rho - 1) \cdot C \cdot T \tag{3.38}$$

By definition, the number of vehicles in the system $L_s(t)$ (See Sect. 1.3) is

$$L_s(t) = L_c(t) + 1 \tag{3.39}$$

With Eq. (3.38), Eq. (3.39) becomes

$$L_s(t) = L_{s0} + (\rho - 1) \cdot C \cdot T \tag{3.40}$$

$L_s(t_0) = L_{s0} = L_{c0} + 1$ be the number of users in the system at the initial instant t_0.

The calculation of the average waiting time in the queue $\overline{w}_c$ and of the time spent in the system $\overline{w}_s$ for the vehicles that arrived between t_0 and $t = t_0 + T$ is straightforward, i.e., the last vehicle in a queue with a length of L_{c0} waits in the queue for a time $w_c(t_0)$ that, on the basis of Eq. (3.29), is equal to the segment $\overline{bf}$ in Fig. 3.8:

$$w_c(t_0) = \overline{bf} = \frac{L_{co}}{C} \tag{3.41}$$

Again, with Eq. (3.29), the waiting time in the queue $w_c(t)$ of the last vehicle that reach the queue at $t = t_0 + T$ is (See Fig. 3.8)

$$w_c(t) = \overline{cd} = \frac{L_c(t)}{C} \tag{3.42}$$

and therefore, for Eq. (3.38),

$$w_c(t) = \frac{L_{c0}}{C} + (\rho - 1) \cdot T \tag{3.43}$$

The mean $\overline{w}_c$ (See Sect. 1.3, Eq. (1.22)) is

$$\overline{w}_c = \frac{w_c(t_0) + w_c(t)}{2} \tag{3.44}$$

then, with Eqs. (3.41) and (3.43),

$$\overline{w}_c = \frac{1}{2}\left[\frac{L_{c0}}{C} + \frac{L_{c0}}{C} + (\rho - 1) \cdot T\right] = \frac{L_{c0}}{C} + \frac{(\rho - 1) \cdot T}{2} \tag{3.45}$$

The average value $\overline{w}_c$ is equal to the value of the waiting time of the vehicle that arrives in the middle of the interval T.

Average value $\overline{w}_c$ can also be obtained on the basis of Eq. (3.30) as a ratio between the area S(bcdfb) and the number of vehicles that arrived in the waiting queue between t_0 and $t = t_0 + T$:

$$\overline{w}_c = \frac{S(bcdfb)}{Q_e \cdot T} \tag{3.46}$$

Since the relationship between the average waiting time in the queue $\overline{w}_c$ and the time spent in the system $\overline{w}_s$ is expressed by Eq. (3.4), since the number of users in the system is $L_{s0} = L_{c0} + 1$ (Eq. (3.39) with $t = t_0$), from Eq. (3.45), we have:

$$\overline{w}_s = \frac{L_{c0}}{C} + \frac{(\rho - 1) \cdot T}{2} + \frac{1}{C} = \frac{L_{s0}}{C} + \frac{(\rho - 1) \cdot T}{2} \tag{3.47}$$

To give an example of the application of the relationship obtained, assume that $L_{c0} = 4$ veh at t_0.

Starting with t_0, we also obtain $Q_e = 1100$ veh/h $= 0.306$ veh/s, $C = 980$ veh/h $= 0.272$ veh/s, and, therefore, $\rho = Q_e/C = 1.125$.

Now, we want to calculate the maximum length of the queue after $T = 10$ min $= 600$ s, the average waiting time in the queue $\overline{w}_c$, and the time spent in the system $\overline{w}_s$ for the vehicles that arrived during the peak period T.

With Eq. (3.38) we obtain

$$L_c(t) = L_{c0} + (\rho - 1) \cdot C \cdot T = 4 + (1.125 - 1) \cdot 0.272 \cdot 600 = 24.4 \text{ veh}$$

From Eq. (3.45)

$$\overline{w}_c = \frac{L_{c0}}{C} + \frac{(\rho - 1) \cdot T}{2} = \frac{4}{0.272} + \frac{(1.125 - 1) \cdot 600}{2} = 52.21 \text{ s}$$

From Eq. (3.47)

$$\overline{w}_s = \frac{L_{s0}}{C} + \frac{(\rho - 1) \cdot T}{2} = \frac{5}{0.272} + \frac{(1.125 - 1) \cdot 600}{2} = 55.88 \text{ s}$$

For Eq. (3.4), the difference $\overline{w}_s - \overline{w}_c$ must be equal to the service time $T_s = 1/C = 3.67$ s; in fact, we have $\overline{w}_s - \overline{w}_c = 55.88 - 52.21 = 3.67$ s.

The just obtained value $\overline{w}_c$ can be also calculated with the evaluation of the area S(bcdfb) (See Fig. 3.8) by applying Eq. (3.46). We have

$$S(\text{bcdfb}) = \frac{1}{2} \cdot \left[\frac{L_{c0}}{C} + \frac{L_c(t)}{C}\right] \cdot \left[\left(T - \frac{L_{c0}}{C}\right) + T_{cd}\right] \cdot C =$$

$$= \frac{1}{2} \cdot \left[\frac{4}{0.272} + \frac{24.4}{0.272}\right] \cdot \left[\left(600 - \frac{4}{0.272}\right) + 89.71\right] \cdot 0.272 =$$

$$= \frac{1}{2} \cdot [14.71 + 89.71] \cdot [(600 - 14.71) + 89.71] \cdot 0.272 \cong 9585 \text{ veh} \cdot \text{s}$$

and, therefore, with Eq. (3.46)

$$\overline{w}_c = \frac{S(\text{bcdefb})}{Q_e \cdot T} = \frac{9585}{0.306 \cdot 600} = 52.21 \text{ s}$$

Sometimes [4], the average waiting times $\overline{w}_c$ and $\overline{w}_s$ are expressed in a slightly different way from Eqs. (3.45) and (3.47), because L_{c0} and L_{s0} are meant to be the number of vehicles that a user who arrives at the initial instant t_0 finds already in the queue or in the system, respectively. Under this assumption, the initial length of the queue $L_c(t_0)$ is then equal to $L_c(t_0) = L_{c0} + 1$; similarly, for $L_s(t_0)$, the result is $L_s(t_0) = L_{s0} + 1$. Repeating the deductive procedure that led to Eqs. (3.45) and (3.47), we can easily obtain

$$\overline{w}_c = \frac{L_{c0} + 1}{C} + \frac{(\rho - 1) \cdot T}{2} \tag{3.48}$$

$$\overline{w}_s = \frac{L_{s0} + 1}{C} + \frac{(\rho - 1) \cdot T}{2} \tag{3.49}$$

3.2.2.2 Second Remarkable Case and Worked Examples

With the criteria presented so far, it is also easy to calculate the variables of the waiting phenomenon associated with a traffic demand and capacity behavior of the type shown in Fig. 3.9, i.e., at the end of the peak period T (i.e., the instant $t_p = t_0 + T$), the entry oversaturation ends and the result is, starting with t_p, systematically $Q_{e1} < C_1$ and $C_1 > C$.

For Eqs. (3.24) and (3.27), the cumulative arrival count A(t) is the line (bcd); the cumulative departure count D(t) is the line (afd).

The maximum queue length L_{cmax} and the maximum number of vehicles in the system L_{smax} are obtained at the end of the traffic peak (at $t_p = t_0 + T$) by means of Eqs. (3.38) and (3.40), respectively.

Figure 3.9 shows how the waiting phenomenon continues, diminishing little by little after the peak period T, for another interval T_d.

Only at the end of T_d (at the instant t_d), the queue is completely cleared, and we can write $L_c(t_d) = A(t_d) - D(t_d) = 0$. T_d is obtained from the relationship (See Fig. 3.9)

$$T_d = \frac{L_{cmax}}{C_1 - Q_{e1}} \tag{3.50}$$

From Eqs. (3.38) and (3.1), Eq. (3.50) can be written

$$T_d = \frac{L_{c0} + (\rho - 1) \cdot C \cdot T}{(1 - \rho_1) \cdot C_1} \tag{3.51}$$

To calculate the waiting time $w_c(t_0) = w_{c0} = \overline{bg}$, Eq. (3.41) is valid, while $w_c(t_p) = w_{cmax}$ can be easily obtained from Fig. 3.9.

$$w_{c\,max} = \overline{ce} = \frac{L_{c\,max}}{C_1} \tag{3.52}$$

and, therefore, with Eq. (3.38)

$$w_{c\,max} = \frac{L_{c0} + (\rho - 1) \cdot C \cdot T}{C_1} \tag{3.53}$$

To calculate the average waiting time $\overline{w}_c$ for the vehicles that arrived during T, Eq. (3.30) is used.

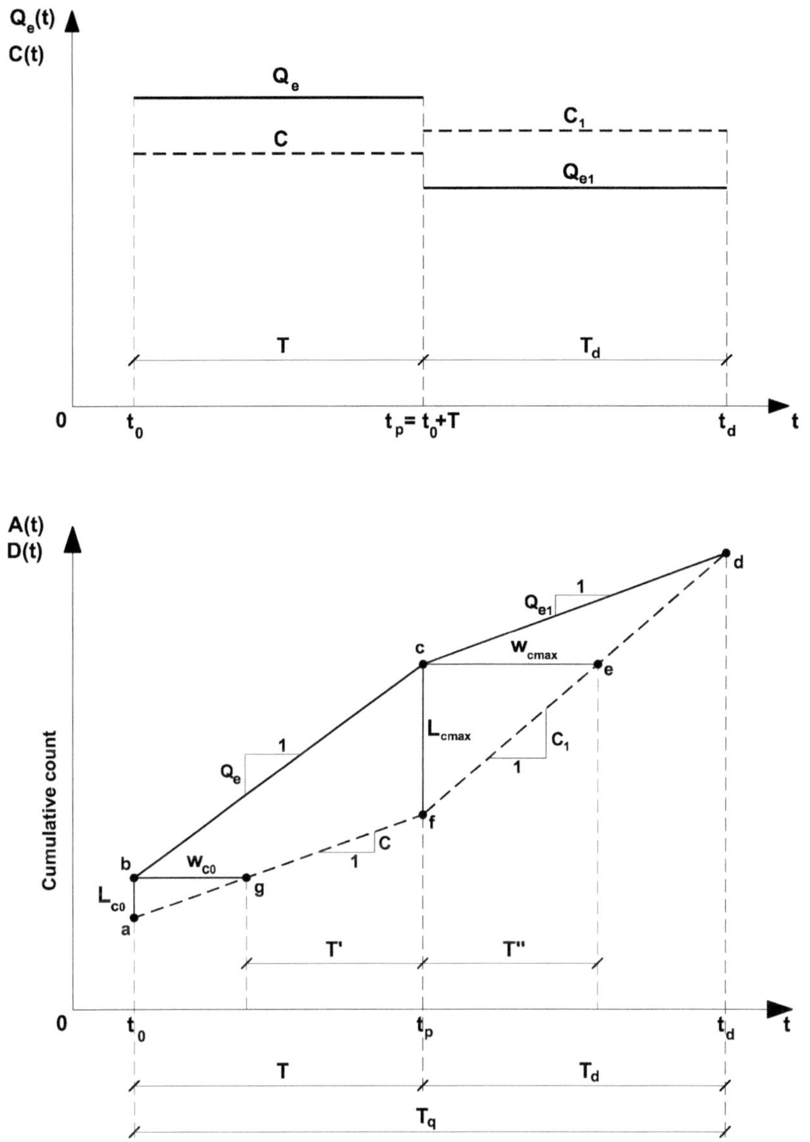

Fig. 3.9 Traffic demand, capacity, and cumulative arrival and departure values for the second remarkable case

In this case, the total waiting time W_c is equal to the area S(bcfgb):

$$W_c = \frac{L_{cmax} \cdot w_{cmax}}{2} + \frac{(L_{cmax} + L_{c0}) \cdot T}{2} - \frac{1}{2} \cdot L_{c0} \cdot \frac{L_{c0}}{C} = \\ = \frac{1}{2} \cdot (A + B - D) \tag{3.54}$$

where, expressing the terms of the second member of Eq. (3.54) with Eqs. (3.38) and (3.53), A, B, and D are

$$A = \frac{[L_{c0} + (\rho - 1) \cdot C \cdot T]^2}{C_1} \tag{3.55}$$

$$B = 2 \cdot L_{c0} \cdot T + (\rho - 1) \cdot C \cdot T^2 \tag{3.56}$$

$$D = \frac{L_{c0}^2}{C} \tag{3.57}$$

Having put

$$E = Q_e \cdot T \tag{3.58}$$

with Eq. (3.30) we obtain

$$\overline{w}_c = \frac{1}{2 \cdot E} \cdot (A + B - D) \tag{3.59}$$

To calculate the average waiting time in the queue $\overline{w}_{cq}$ relative to all the delayed vehicles, i.e., vehicles that arrived in the queue during the interval $T_q = T + T_d$, we must take into account the total waiting time in the queueW_c^*, which is equal to the area S(bcdfgb). Thus we have (See Fig. 3.9)

$$W_c^* = \frac{L_{cmax} \cdot T_d}{2} + \frac{(L_{cmax} + L_{c0}) \cdot T}{2} - \frac{1}{2} \cdot L_{c0} \cdot \frac{L_{c0}}{C} \tag{3.60}$$

Equation (3.30), with Eqs. (3.38) and (3.51), is written

$$W_c^* = \frac{1}{2} \cdot (A^* + B - D) \tag{3.61}$$

where

$$A^* = \frac{[L_{c0} + (\rho - 1) \cdot C \cdot T]^2}{(1 - \rho_1) \cdot C_1} \tag{3.62}$$

while B and D are given by Eqs. (3.56) and (3.57), respectively.

If we put now

$$E^* = Q_e \cdot T + Q_{e1} \cdot T_d \tag{3.63}$$

that, for Eq. (3.51), becomes

$$E^* = Q_e \cdot T + Q_{e1} \cdot \frac{L_{c0} + (\rho - 1) \cdot C \cdot T}{(1 - \rho_1) \cdot C_1} \tag{3.64}$$

on the basis of Eq. (3.30), $\overline{w}_{cq}$ is

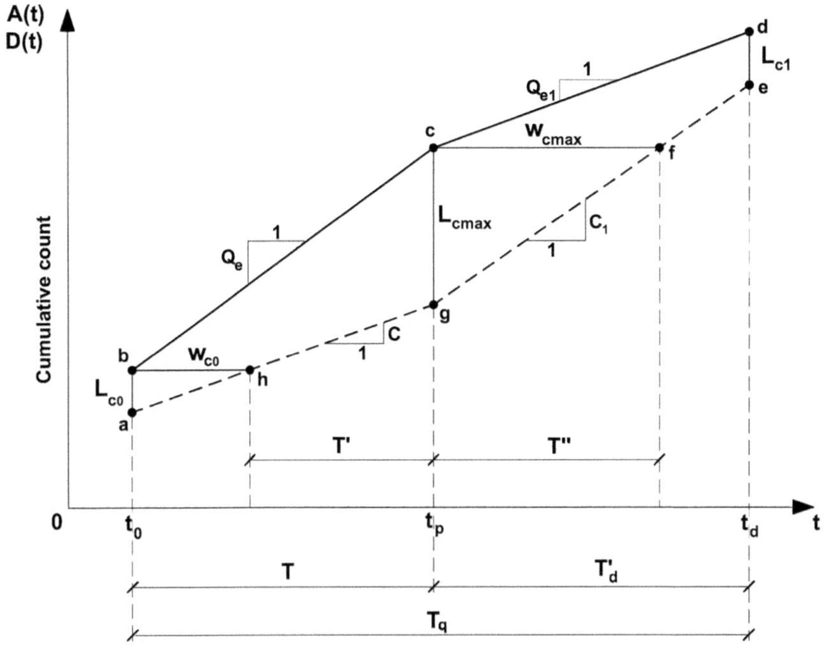

Fig. 3.10 Discharge process of peak traffic effects with the queue final length L_{c1} greater than zero

$$\overline{w}_{cq} = \frac{1}{2 \cdot E^*} \cdot (A^* + B - D) \tag{3.65}$$

If the effects of the peak period are considered to end when L_{c1} vehicles are present in the queue (See Fig. 3.10), Eq. (3.50) of T_d is modified as

$$T'_d = \frac{L_{cmax} - L_{c1}}{C_1 - Q_{e1}} = \frac{L_{cmax} - L_{c1}}{C_1 \cdot (1 - \rho_1)} \tag{3.66}$$

because, by Eq. (3.1), $\rho_1 = Q_{e1}/C_1$.

The relationship that gives the total waiting time W_c in this case is written taking into account the area S(bcdeghb) (See Fig. 3.10), and it is

$$W_c^{**} = \frac{1}{2} \cdot (A^{**} + B - D) \tag{3.67}$$

where

$$A^{**} = \frac{[L_{c0} + (\rho - 1) \cdot C \cdot T]^2 - L_{c1}^2}{(1 - \rho_1) \cdot C_1} \tag{3.68}$$

since L_{cmax} is always given by Eq. (3.38) and T'_d by Eq. (3.66); again, B and D are given by Eqs. (3.56) and (3.57), respectively.

Putting

$$E^{**} = Q_e \cdot T + Q_{e1} \cdot T_d' = Q_e \cdot T + Q_{e1} \cdot \frac{[L_{c0} + (\rho - 1) \cdot C \cdot T] - L_{c1}}{(1 - \rho_1) \cdot C_1} \tag{3.69}$$

with Eq. (3.69), for Eq. (3.30), the average waiting time in the queue $\overline{w}'_{cq}$ is

$$\overline{w}'_{cq} = \frac{1}{2 \cdot E^{**}} \cdot (A^{**} + B - D) \tag{3.70}$$

where B and D are again given by Eqs. (3.56) and (3.57), respectively.

Finally, the calculation of the average time spent in the system $\overline{w}_s$ for vehicles that arrived during the peak T is conducted, on the basis of Eq. (3.31), adding the mean value $\overline{T}_s$ of the service time T_s to the mean $\overline{w}_c$.

$\overline{T}_s$ is evaluated taking into account that (See Fig. 3.9) during the fraction T' of T, $T_s = 1/C$, while in T'', $T_s = 1/C_1$.

Therefore, for $\overline{T}_s$ as the weighted mean, we have

$$\overline{T}_s = \frac{\frac{1}{C} \cdot C \cdot T' + \frac{1}{C_1} \cdot C_1 \cdot T''}{C \cdot T' + C_1 \cdot T''} = \frac{T' + T''}{C \cdot T' + C_1 \cdot T''} \tag{3.71}$$

where

$$T' = T - (L_{c0}/C) \tag{3.72}$$

$$T'' = w_{cmax} = L_{cmax}/C_1 \tag{3.73}$$

In conclusion, for Eq. (3.31) we have, with Eqs. (3.59) and (3.71)

$$\overline{w}_s = \overline{T}_s + \overline{w}_c = \frac{T' + T''}{C \cdot T' + C_1 \cdot T''} + \frac{1}{2 \cdot E}(A + B - D) \tag{3.74}$$

With the same procedure that led to Eq. (3.71), we can obtain the mean values $\overline{T}_{sq}$ and $\overline{T}'_{sq}$ of service time relative to all the delayed vehicles, in the case of complete clearance of the queue at the end of the interval T_d (Eq. (3.50)) and in the case of a queue L_{c1} greater than zero at the end of T'_d (Eq. (3.66)), respectively.

Thus, we have the following relationship:

$$\overline{T}_{sq} = \frac{\frac{1}{C} \cdot C \cdot T' + \frac{1}{C_1} \cdot C_1 \cdot T_d}{C \cdot T' + C_1 \cdot T_d} = \frac{T' + T_d}{C \cdot T' + C_1 \cdot T_d} \tag{3.75}$$

where T' and T_d are given by Eqs. (3.72) and (3.50), respectively.

$$\overline{T'}_{sq} = \frac{\frac{1}{C} \cdot C \cdot T' + \frac{1}{C_1} \cdot C_1 \cdot T'_d}{C \cdot T' + C_1 \cdot T'_d} = \frac{T' + T'_d}{C \cdot T' + C_1 \cdot T'_d} \tag{3.76}$$

where T' and T'_d are given by Eqs. (3.72) and (3.66), respectively.

In conclusion, on the basis of Eq. (3.31), with Eqs. (3.65) and (3.75), we have the average time spent in the system

$$\overline{w}_{sq} = \overline{T}_{sq} + \overline{w}_{cq} = \frac{T' + T_d}{C \cdot T' + C_1 \cdot T_d} + \frac{1}{2 \cdot E^*}(A^* + B - D) \tag{3.77}$$

Similarly, again on the basis of Eq. (3.31), with Eqs. (3.70) and (3.76), we obtain:

$$\overline{w}'_{sq} = \overline{T}'_{sq} + \overline{w}'_{cq} = \frac{T' + T'_d}{C \cdot T' + C_1 \cdot T'_d} + \frac{1}{2 \cdot E^{**}}(A^{**} + B - D) \tag{3.78}$$

As an example, we consider the case of an entry for which there are $L_{c0} = 4$ veh queuing at time t_0. We refer here to Fig. 3.9.

In $T = t_p - t_0 = 15$ min $= 900$ s, for the arrival demand and capacity we have, $Q_e = 1100$ veh/h $= 0.306$ veh/s and $C = 980$ veh/h $= 0.272$ veh/s, respectively, and, therefore, $\rho = Q_e/C = 1.125$.

Starting with t_p, we have $Q_{e1} = 850$ veh/h $= 0.236$ veh/s and $C_1 = 1010$ veh/h $=$ 0.281 veh/s, and, therefore, $\rho_1 = Q_{e\,1}/\,C_1 = 0.840$.

At the end of the peak period T, the maximum queue length L_{cmax} (Eq. (3.38)) is recorded

$$L_{c\,max} = L_{c0} + (\rho - 1) \cdot C \cdot T = 4 + (1.125 - 1) \cdot 0.272 \cdot 900 = 34.6\,\text{veh}$$

and the maximum waiting time in the queue w_{cmax} (Eq. (3.52)) is

$$w_{c\,max} = \frac{L_{c\,max}}{C_1} = \frac{34.6}{0.281} = 123.13\text{s} \cong 2\,\text{min}$$

The average waiting time in the queue $\overline{w}_c$ for the vehicles that arrived during the peak period T is evaluated by calculating, with Eqs. (3.55), (3.56), and (3.57), the quantities A, B, and D, which must be inserted into Eq. (3.59):

$$A = \frac{[L_{c\,max}]^2}{C_1} = \frac{34.6^2}{0.281} = 4260.36\,\text{veh} \cdot \text{s}$$

$$B = 2 \cdot L_{c0} \cdot T + (\rho - 1) \cdot C \cdot T^2 = 2 \cdot 4 \cdot 900 + (1.125 - 1) \cdot 0.272 \cdot 900^2 = 34740\,\text{veh} \cdot \text{s}$$

$$\mathrm{D} = \frac{\mathrm{L}_{c0}^{2}}{\mathrm{C}} = \frac{4^2}{0.272} = 58.82\,\mathrm{veh}\cdot\mathrm{s}$$

$$\overline{\mathrm{w}}_{\mathrm{c}} = \frac{1}{2\cdot\mathrm{Q_e}\cdot T}\cdot(\mathrm{A}+\mathrm{B}-\mathrm{D}) = \frac{1}{2\cdot 0.306\cdot 900}\cdot(4260.36+34740-58.82) =$$
$$= 70.70\mathrm{s} = 1.18\mathrm{min}$$

The average time spent in the system relative to the vehicles that arrived during the traffic peak, $\overline{\mathrm{w}}_{\mathrm{s}}$, is obtained with Eq. (3.74), once that by Eqs. (3.72) and (3.73) the periods T' and T'' are evaluated and, therefore, with them, according to Eq. (3.71), the mean service time $\overline{\mathrm{T}}_{\mathrm{s}}$

$$\mathrm{T}' = \mathrm{T} - (\mathrm{L}_{c0}/\mathrm{C}) = 900 - (4/0.272) = 885.3\,\mathrm{s}$$

$$\mathrm{T}'' = \mathrm{w}_{\mathrm{c\,max}} = 123.13\,\mathrm{s}$$

$$\overline{\mathrm{T}}_s = \frac{\mathrm{T}'+\mathrm{T}''}{\mathrm{C}\cdot\mathrm{T}'+\mathrm{C}_1\cdot\mathrm{T}''} = \frac{885.29+123.13}{0.272\cdot 885.29+0.281\cdot 123.13} = 3.66\,\mathrm{s}$$

$$\overline{\mathrm{w}}_{\mathrm{s}} = \overline{\mathrm{T}}_s + \overline{\mathrm{w}}_c = 3.66+70.70 = 74.36\,\mathrm{s}$$

The duration of the interval $\mathrm{T_d}$, where the effects are visible once the peak ends, by Eq. (3.50), is

$$\mathrm{T_d} = \frac{\mathrm{L}_{\mathrm{c\,max}}}{\mathrm{C}_1-\mathrm{Q}_{\mathrm{e}1}} = \frac{34.6}{0.281-0.236} = 768.89\,\mathrm{s}$$

Having determined the value of $\mathrm{T_d}$, it is possible to calculate E^* (Eq. (3.63)) and, once A^* has been evaluated by Eq. (3.62), Eq. (3.65) can be used to obtain the average waiting time in the queue $\overline{\mathrm{w}}_{\mathrm{cq}}$ relative to all the vehicles that arrived during the period ($\mathrm{T_q} = \mathrm{T} + \mathrm{T_d}$), (See Fig. 3.9).

Therefore, we have

$$\mathrm{E}^* = \mathrm{Q_e}\cdot\mathrm{T}+\mathrm{Q}_{\mathrm{e}1}\cdot\mathrm{T_d} = 0.306\cdot 900+0.236\cdot 768.89 = 456.86\,\mathrm{veh}$$

$$\mathrm{A}^* = \frac{[\mathrm{L}_{\mathrm{c\,max}}]^2}{(1-\rho_1)\cdot\mathrm{C}_1} = \frac{34.6^2}{(1-842)\cdot 0.281} = 26627.22\,\mathrm{veh}\cdot\mathrm{s}$$

and by them, with B and D calculated above,

$$\overline{\mathrm{w}}_{\mathrm{cq}} = \frac{1}{2\cdot\mathrm{E}^*}\cdot(\mathrm{A}^*+\mathrm{B}-\mathrm{D}) = \frac{1}{2\cdot 456.86}\cdot(26627.22+34740-58.82) =$$
$$= 67.10\,\mathrm{s} = 1.12\,\mathrm{min}$$

The average time spent in the system $\overline{w}_{sq}$ relative to the vehicles that arrived during T_q is given by Eq. (3.77), once known (with $T' = 885.29$ s and $T_d = 768.89$ s), the mean service time $\overline{T}_{sq}$ (Eq. (3.75))

$$\overline{T}_{sq} = \frac{T' + T_d}{C \cdot T' + C_1 \cdot T_d} = \frac{885.29 + 768.89}{0.272 \cdot 885.29 + 0.280 \cdot 768.89} = 3.62\,\text{s}$$

Then we have

$$\overline{w}_{sq} = \overline{T}_{sq} + \overline{w}_{cq} = 3.62 + 67.10 = 70.72\ \text{s} = 1.18\,\text{min}$$

If we consider that the effects of the peak period ends in presence of a queue L_{c1} at the end of the period T'_d (which is still affected by the demand peak in T, Fig. 3.10), the extension of T'_d is obtained by Eq. (3.66). If we assume $L_{c1} = 3$ veh and use the data from this example, we obtain

$$T'_d = \frac{L_{c\,max} - L_{c1}}{C_1 - Q_{e1}} = \frac{34.6 - 3}{0.281 - 0.236} = 702.22\,\text{s}$$

The calculation of the average waiting time in the queues $\overline{w}'_{cq}$ and $\overline{w}'_{sq}$ requires the evaluation of A^{**} (Eq. (3.68)), E^{**} (Eq. (3.69)), and $\overline{T}'_{sq}$ (Eq. (3.76)).

With the above-mentioned relationships, we have

$$A^{**} = \frac{L^2_{c\,max} - L^2_{c1}}{(1 - \rho_1) \cdot C_1} = \frac{34.6^2 - 3^2}{(1 - 0.840) \cdot 0.281} = 26427.05\ \text{veh} \cdot \text{s}$$

$$E^{**} = Q_e \cdot T + Q_{e1} \cdot T'_d = 0.306 \cdot 900 + 0.236 \cdot 702.22 = 441.12\,\text{veh}$$

$$\overline{T}'_{sq} = \frac{T' + T'_d}{C \cdot T' + C_1 \cdot T'_d} = \frac{885.29 + 702.22}{0.272 \cdot 885.29 + 0.281 \cdot 702.22} = 3.62\,\text{s}$$

with B and D already calculated, by Eq. (3.70), we obtain

$$\overline{w}'_{cq} = \tfrac{1}{2 \cdot E^{**}} \cdot (A^{**} + B - D) = \tfrac{1}{2 \cdot 441.12} \cdot (26427.05 + 34740 - 58.82) =$$
$$= 69.26\ \text{s} = 1.15\,\text{min}$$

Knowing $\overline{w}'_{cq}$ and $\overline{T}'_{sq}$, from Eq. (3.78) we obtain

$$\overline{w}'_{sq} = \overline{T}'_{sq} + \overline{w}'_{cq} = 3.62 + 69.26 = 72.88\,\text{s} = 1.21\ \text{min}$$

3.2.3 Further Expressions of the Average Waiting Times

The expressions of the mean times $\overline{w}_c$, $\overline{w}_{cq}$, and $\overline{w}'_{cq}$ given in the previous sections were obtained from the waiting times only for the vehicles that arrived during the period T and in the following period that was still affected by the peak demand.

Now, if we want to take into account all the vehicles included in the queue starting with the beginning t_0 of the peak period, it is obvious that the term D in Eqs. (59), (65), and (70) must not be subtracted from the terms A and B in brackets. In addition, the values of E, E^*, and E^{**} must be added to the value L_{c0} of the length of the possible queue that may be present at the entry at the instant t_0. Thus, we have

$$\overline{w}_c = \frac{1}{2 \cdot (E + L_{c0})} \cdot (A + B) \tag{3.79}$$

$$\overline{w}_{cq} = \frac{1}{2 \cdot (E^* + L_{c0})} \cdot (A^* + B) \tag{3.80}$$

$$\overline{w}'_{cq} = \frac{1}{2 \cdot (E^{**} + L_{c0})} \cdot (A^{**} + B) \tag{3.81}$$

Even for the expressions of times spent in the system, Eqs. (3.74), (3.77), and (3.78) are modified to take into account the possible presence of the queue at the instant t_0.

Also, the means of the service times must, in fact, be added to Eqs. (3.79), (3.80), and (3.81). In the expressions (3.71), (3.75) and (3.76) of the above-mentioned means, the interval T′ must be substituted for total T (See Fig. 3.10) to account for the time (T–T′) used to serve the vehicles that form L_{c0}:

$$\overline{w}_s = \overline{T}_s + \overline{w}_c = \frac{T + T''}{C \cdot T + C_1 \cdot T''} + \frac{1}{2 \cdot (E + L_{co})}(A + B) \tag{3.82}$$

$$\overline{w}_{sq} = \overline{T}_{sq} + \overline{w}_{cq} = \frac{T + T_d}{C \cdot T + C_1 \cdot T_d} + \frac{1}{2 \cdot (E^* + L_{c0})}(A^* + B) \tag{3.83}$$

$$\overline{w}'_{sq} = \overline{T}'_{sq} + \overline{w}'_{cq} = \frac{T + T'_d}{C \cdot T + C_1 \cdot T'_d} + \frac{1}{2 \cdot (E^{**} + L_{c0})}(A^{**} + B) \tag{3.84}$$

Frequently [5], as indicators of the effects of traffic demand peaks, average conventional values of waiting times are used (in the queue or in the system) obtained as a ratio between total delays

$$W_c^* = \frac{1}{2}(A^* + B) \tag{3.85}$$

$$W_c^{**} = \frac{1}{2}(A^{**} + B) \tag{3.86}$$

and only the vehicles that arrived during the peak period T, equal to $Q_e \cdot T$.

Using the same criterion, the mean service time must be calculated only in relation to the above-mentioned vehicles. Thus, we have

$$\overline{T}_{sq} = \overline{T}'_{sq} = \frac{T + T''}{C \cdot T + C_1 \cdot T''} \tag{3.87}$$

In conclusion, by Eqs. (3.85), (3.86), and (3.87), we have

$$\overline{w}_{cq} = \frac{1}{2 \cdot Q_e \cdot T} \cdot (A^* + B) \tag{3.88}$$

$$\overline{w}'_{cq} = \frac{1}{2 \cdot Q_e \cdot T} \cdot (A^{**} + B) \tag{3.89}$$

$$\overline{w}_{sq} = \overline{T}_{sq} + \overline{w}_{cq} = \frac{T + T''}{C \cdot T + C_1 \cdot T''} + \frac{1}{2 \cdot Q_e \cdot T} \cdot (A^* + B) \tag{3.90}$$

$$\overline{w}'_{sq} = \overline{T}'_{sq} + \overline{w}'_{cq} = \frac{T + T''}{C \cdot T + C_1 \cdot T''} + \frac{1}{2 \cdot Q_e \cdot T} \cdot (A^{**} + B) \tag{3.91}$$

3.2.4 Concluding Remarks

The results illustrated in the previous sections were based on the following assumptions:

- the cumulative arrival A(t) and departures D(t) counts are generally approximated with a continuous process (fluid approximation) that, in the case of a succession of periods with constant traffic demand and capacity, are continuous at intervals (piecewise-linear);
- during a period of constant traffic demand Q_e, the vehicles reach the leg at constant intervent times $\tau_A = 1/Q_e$;
- during a period when capacity C is constant, the vehicles depart from the queue (or the system) at constant time rates equal to a $\tau_D = 1/C$.

The fluid approximation becomes even more valid when the system is congested.

We recall that congestion is recorded if the entry is oversaturated, saturated, or undersaturated but starting with an initial state with a long queue.

During the congestion, the queue length is greater than unity, and the waiting times are greater than the mean service time. Under these conditions, the values of the discontinuities of A(t) and D(t) (equal to unity when there is an arrival or a departure – Fig. 3.5) are small with respect to the mean values in the time of the same functions, i.e., it follows that we have, for A(t) and D(t), small increases relative to the progression of the processes. Similarly, we have small differences between the continuous approximations and the starting relationships (step functions).

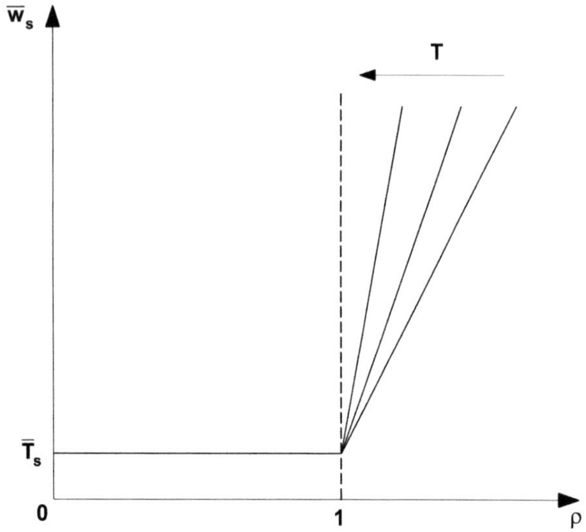

Fig. 3.11 Generic behavior of the average time spent in the system versus the value of saturation degree

The regularity assumptions of the arrival and departure processes become even more plausible when the traffic conditions become heavier, since, at these conditions, the random changes of the above-mentioned processes tend to diminish, with the consequence that the prevailing effects are given by the means, while the fluctuations around them tend to diminish. [4]

Finally, it is worth noting that the solutions provided by the deterministic approach belong to the type of time-dependent solutions of the queuing theory. In fact, unlike the probabilistic solutions, they take into account both the finite extension of the congestion periods of the system and the prolongation of peak effects.

As an example of a time-dependent solution, Fig. 3.11 shows the generic behavior of the average time spent in the system (provided by Eq. (3.49)) as the peak interval T value increases when $L_{s0} = 0$.

We will present further time-dependent solutions (not deterministic) in the following section.

3.3 Time-Dependent Solutions and Waiting Phenomena

As we have already pointed out in the previous sections, the solutions provided by the probabilistic approach are valid at a steady-state condition.

[4] In queuing theory, the fluid approximation (deterministic approximation) has a wider meaning than the one just recalled, since it involves the consideration of the relationship between the arrival and departure process realizations and the respective levels (means affected by the previous values) and the mean time values. About this aspect, see [3] and [7].

For a real system, these conditions are only partially possible.

The above-mentioned solutions are considered to be acceptable approximations for the entries "i" if the time T_i during which traffic demands Q_{ei} and capacity C_i may be assumed to be constant is sufficiently wide and if the entry is undersaturated in T_i and the traffic intensity $\rho_i = Q_{ei}/C_i$ is suitably smaller than unity. [5]

In conclusion, by the probabilistic results it is not possible to study situations in which Q_{ei} is time-dependent, and it can also exceed capacity C_i in a specific peak period.

Regarding the deterministic approach, in the observations presented in previous Sect. 3.2.4, we pointed out that the results become more reliable as the entry becomes more oversaturated (in presence or absence of queues at the initial instant of system observation), i.e., with the increase of $\rho_i > 1$. In fact, under this circumstance, the random effects on the state variables (queues and waiting times) become smaller, and the expected values of the same stochastic state variables near deterministic means.

In other words, for the conditions at which the entry is oversaturated with $\rho_i > 1$ but not with $\rho_i >> 1$, the deterministic approach cannot take into account the random effects of the evolution of the waiting phenomenon in the real system, leading to the systematic underestimation of the values of the state variables realizations.

Then, to describe the non-steady situations, i.e., variability Q_{ei} and/or oversaturation of the entry, but with ρ_i not sufficiently greater than unity, the two approaches – probabilistic and deterministic – so far described are not suitable.

In mathematical queuing theory, thorough discussions are available for these cases. Some of them are capable of solving particular problems in a precise way, others in approximate ways, but none of them is simple enough for practical applications [3], [6], [7].

That is why, in the analysis of a system under steady-state conditions, we generally use heuristic criteria of transition from statistical solutions to deterministic solutions. Thus, we obtain (for the calculation of queue lengths, number of users in the system, and waiting times) relationships in which traffic intensity ρ and the time T relative to the duration of system observation or the peak period appear as independent variables.

More precisely, a state variable δ (e.g., L_c, L_s, w_c, or w_s) is calculated as average $<\delta>$ for a fixed value of the interval T, suitably combining its expected value $E[\delta]$ under statistical equilibrium conditions (steady-state conditions) with its mean deterministic value $\bar{\delta}$, so that (See Fig. 3.12) the oblique asymptote of the relationship $<\delta> = <\delta(\rho)>$ coincides with the deterministic function that yields $\bar{\delta}$.

The determination of $<\delta> = <\delta(\rho)>$ may be performed in different ways.

The two following criteria are usually used.

[5] For the value of T_i, see the indication included in footnote 4 of Chap. 1 and the following Chap. 4.

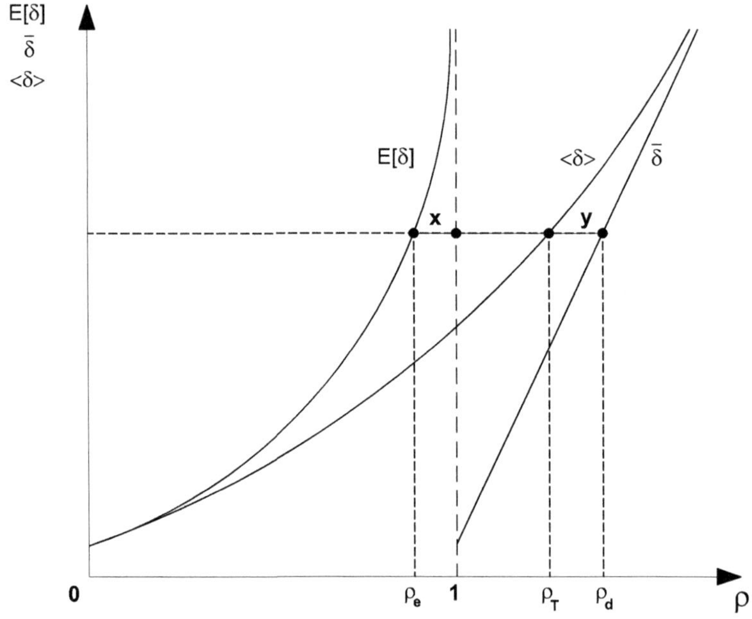

Fig. 3.12 Averages of $E[\delta]$, $\bar{\delta}$, and $<\delta>$ versus traffic intensity ρ

Given generic equal values of $E[\delta]$, $\bar{\delta}$, and $<\delta>$ ($E[\delta] = \delta = <\delta>$), let x be the distance between $E[\delta]$ and the vertical asymptote for $\rho = 1$, and let y be the distance between the curve $<\delta> = <\delta(\rho)>$ (to be determined) and the half-line $\bar{\delta} = \bar{\delta}(\rho)$ that is its oblique asymptote.

Then, we can write

$$x = y \tag{3.92}$$

or, alternatively,

$$x{:}1 = y{:}\rho_d \tag{3.93}$$

In terms of traffic intensity, for Eq. (3.92), we have (See Fig. 3.12)

$$1 - \rho_e = \rho_d - \rho_T \tag{3.94}$$

thus

$$\rho_e = \rho_T - (\rho_d - 1) \tag{3.95}$$

If we assume, instead, Eq. (3.93) we have

$$1 - \rho_e = \frac{\rho_d - \rho_T}{\rho_d} \tag{3.96}$$

thus

$$\rho_e = \frac{\rho_T}{\rho_d} \tag{3.97}$$

Starting with Eq. (3.95) or Eq. (3.97) and choosing explicit ρ_e and ρ_d values as functions of $E[\delta]$ and $\overline{\delta}$, we have, for an assigned T, two (different) expressions of $<\delta> = <\delta(\rho)>$.

Given the heuristic nature of the deductive criterion of the behavior of $<\delta>$ under time-dependent conditions (i.e., non-steady), there are no specific reasons to prefer one of the Eqs. (3.92) and (3.93) over the other (or other equations that may be used to obtain the transition $<\delta>$ between $E[\delta]$ and $\overline{\delta}$ with an asymptotic behavior to $\overline{\delta}$).

In the following sections, in accordance with the main trend in the technical literature [4, 5], Eq. (3.92) will be used to obtain time-dependent solutions for the average number of users in the system $\overline{L}_s$, the queue length $\overline{L}_c$, and the waiting times $\overline{w}_s$ and $\overline{w}_c$.

3.3.1 Time-Dependent Solutions for the Number of Users in the System and in the Queue. A Worked Example

The behaviors of traffic demand Q_e and capacity C are described in the previous Sect. 3.2.2.1 (first remarkable case) (See Fig. 3.8).

For the average $E[L_s]$ of the number of users in the system under a steady condition, Eq. (3.18) is used

$$E[L_s] = \frac{\rho}{1 - \rho} \tag{3.18}$$

and, for the average $\overline{L}_s$ of the same state variable[6] at the end of the period T of system observation, Eq. (3.40) is used

$$\overline{L}_s(t) = L_{s0} + (\rho - 1) \cdot C \cdot T \tag{3.40}$$

Now, to obtain the transition curve $< L_s >$ between $E[L_s]$ and $\overline{L}_s$, we use Eq. (3.92) or Eq. (3.95)

$$\rho_e = \rho_T - (\rho_d - 1) \tag{3.95}$$

[6] We recall that, in the deterministic approach, the value of a state variable coincides with the average value calculated in the same instant and with the percentiles, regardless of their order.

We recall (See Fig. 3.12) that ρ_e, ρ_d, and ρ_T are the values of traffic intensities that result in

$$E[L_s] = \overline{L}_s = < L_s > \tag{3.98}$$

Therefore, with $\rho = \rho_d$ and Eq. (3.40), we can use Eq. (3.98) to obtain

$$\rho_d = \frac{\overline{L}_s - L_{s0}}{C \cdot T} + 1 = \frac{< L_s > - L_{s0}}{C \cdot T} + 1 \tag{3.99}$$

and with $\rho = \rho_e$ and Eq. (3.18), we can use Eq. (3.98) to obtain

$$< L_s > = \frac{\rho_e}{1 - \rho_e} \tag{3.100}$$

Substituting Eq. (3.99) into Eq. (3.95) we have

$$\rho_e = \rho_T - \frac{< L_s > - L_{s0}}{C \cdot T} \tag{3.101}$$

When ρ_e is provided by Eq. (3.101), from Eq. (3.100) we have

$$< L_s > - \frac{\rho_T - \dfrac{< L_s > - L_{s0}}{C \cdot T}}{1 - \rho_T + \dfrac{< L_s > - L_{s0}}{C \cdot T}} = 0 \tag{3.102}$$

Equation (3.102) provides an implicit definition of the transition curve $< L_s >$ between $E[L_s]$ and $\overline{L}_s$.

Limiting the validity of Eqs. (3.18) and (3.40) to the first quadrant of the plane (ρ; means of L_s), where they have a physical meaning (See Fig. 3.13), from Eq. (3.102), solved with these restrictions, and since ρ_T is a current value of traffic intensity ($\rho = \rho_T$), we have

$$< L_s > = \frac{1}{2}\left(\sqrt{A^2 + B} - A\right) \tag{3.103}$$

with

$$A = (1 - \rho) \cdot C \cdot T + 1 - L_{s0} \tag{3.104}$$

$$B = 4(L_{s0} - \rho \cdot C \cdot T) \tag{3.105}$$

To give an example of the application of Eq. (3.103), consider the following data (See Fig. 3.8): at the initial time t_0, we obtain $L_{s0} = 5$ veh; starting with t_0, $Q_e = 1100$ veh/h $= 0.306$ veh/s, $C = 980$ veh/h $= 0.272$ veh/s, and, thus, $\rho = Q_e/C = 1.125$. The interval of the system observation is assumed to be $T = 10$ min $= 600$ s.

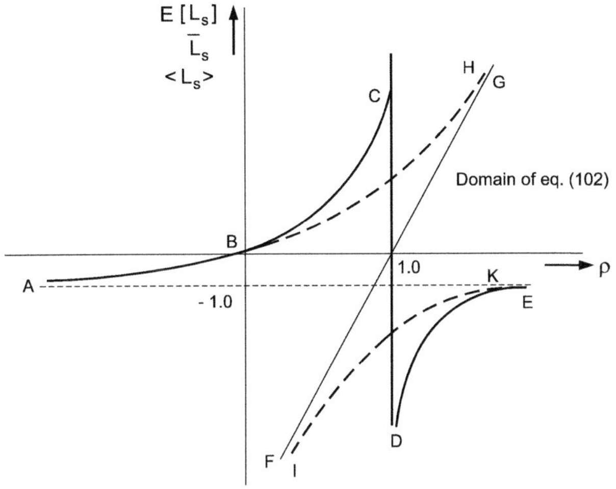

Fig. 3.13 Domains of time-dependent solutions for the number L_s of users in the system

For $< L_s >$ at $t = t_0 + T$, the quantities A and B are evaluated using Eqs. (3.104) and (3.105)

$$A = (1 - \rho) \cdot C \cdot T + 1 - L_{s0} = (1 - 1.125) \cdot 0.272 \cdot 600 + 1 - 5 = -24.4 \text{ veh}$$
$$B = 4 \cdot (L_{s0} + \rho \cdot C \cdot T) = 4 \cdot (5 + 1.125 \cdot 0.272 \cdot 600) = 754.4 \text{ veh}$$

and, therefore, having obtained A and B, we get

$$< L_s > = \frac{1}{2}\left(\sqrt{A^2 + B} - A\right) = \frac{1}{2}\left(\sqrt{(-24.4)^2 + 754.4} + 24.4\right) = 30.57 \text{ veh}$$

Following the same procedure that led to Eq. (3.103), it is easy to deduce the transition relationship between the mean of the queue length $E[L_c]$ under steady-state conditions (See Table 3.3)

$$E[L_c] = \frac{\rho^2}{1 - \rho} \tag{3.106}$$

and the mean deterministic value $\overline{L}_c$ (Eq. (3.38))

$$\overline{L}_c(t) = L_{c0} + (\rho - 1) \cdot C \cdot T \tag{3.38}$$

thus we have for the time-dependent relationship for $< L_c >$

$$< L_c > = \frac{1}{2}\left(\sqrt{D^2 + E} - D\right) \tag{3.107}$$

with

$$D = \frac{(1-\rho)\cdot(C\cdot T)^2 - C\cdot T\cdot L_{c0} + 2\cdot(L_{c0}+\rho\cdot C\cdot T)}{C\cdot T-1} \tag{3.108}$$

$$E = \frac{4\cdot(L_{c0}+\rho\cdot C\cdot T)^2}{C\cdot T-1} \tag{3.109}$$

3.3.2 Time-Dependent Solutions for the Average Time Spent in the System and in the Queue. A Worked Example

We refer here to the behaviors of traffic demand Q_e and capacity C illustrated in Sect. 3.2.2.1 (See Fig. 3.8); we follow the same procedure illustrated in the same section to obtain the expression of $<L_S>$.

Therefore, we use Eq. 3.17 for the expected value $E[w_S]$ of the average time spent in the system under steady-state conditions

$$E[w_S] = \frac{1}{C}\cdot\frac{1}{(1-\rho)} = \frac{1}{C}\cdot\left(1+\frac{\rho}{1-\rho}\right) \tag{3.17}$$

and we use Eq. 3.49 for the deterministic mean of the same state variable[7]

$$\overline{w}_S = \frac{L_{S0}+1}{C} + \frac{(\rho-1)\cdot T}{2} \tag{3.49}$$

Now, we use Eq. (3.93) or Eq. (3.95) to obtain the transition curve $<w_S>$ between $E[w_S]$ and $\overline{w}_S$

$$\rho_e = \rho_T - (\rho_d - 1) \tag{3.95}$$

We recall that (See Fig. 3.12) ρ_e, ρ_d, and ρ_T are values of traffic intensity in correspondence of which it results

$$E[w_S] = \overline{w}_S = <w_S> \tag{3.110}$$

Therefore, with $\rho = \rho_d$ and Eq. (3.49) and using Eq. (3.110), we have

$$\rho_d = \frac{2}{T}\cdot\left(<w_S> - \frac{L_{S0}+1}{C}\right) + 1 \tag{3.111}$$

[7] We recall that $\overline{w}_S$ is the mean of the times spent in the system for vehicles that queue up during the interval T of system observation.

and with $\rho = \rho_e$ and Eq. (3.17) and using Eq. (3.110) again

$$<w_S> = \frac{1}{C} \cdot \left(1 + \frac{\rho_e}{1 - \rho_e}\right) \tag{3.112}$$

Substituting Eq. (3.111) into Eq. (3.95) we have

$$\rho_e = \rho_T + \frac{2}{T} \cdot \left(\frac{L_{s0} + 1}{C} - <w_S>\right) \tag{3.113}$$

Using the value of ρ_e from Eq. (3.113) with Eq. (3.112), we can write

$$<w_S> - \frac{1}{C} \cdot \left(1 + \frac{\rho_T + \frac{2}{T} \cdot \left(\frac{L_{s0}+1}{C} - <w_S>\right)}{1 - \rho_T - \frac{2}{T} \cdot \left(\frac{L_{s0}+1}{C} - <w_S>\right)}\right) = 0 \tag{3.114}$$

Equation (3.114) implicitly defines the transition curve $<w_S>$ between $E[w_S]$ and $\overline{w}_S$. Limiting the validity of Eqs. (3.17) and (3.49) to the first quadrant of the plane (ρ; means of w_S), where they have a physical meaning, we can solve Eq. (3.114) under these restrictions. Since ρ_T is a current value of traffic intensity ($\rho = \rho_T$), the result is

$$<w_S> = \frac{1}{2}\left(\sqrt{J^2 + M} - J\right) \tag{3.115}$$

with

$$J = \frac{T}{2} \cdot (1 - \rho) - \frac{1}{C} \cdot (L_{s0} + 1) \tag{3.116}$$

$$M = \frac{4}{C} \cdot \left[\frac{T}{2} \cdot (1 - \rho) + \frac{1}{2} \cdot \rho \cdot T\right] \tag{3.117}$$

Reusing the data of the worked example given in the previous Sect. 3.3.1, we calculate the time-dependent mean of the average time spent in the system for vehicles that joined the queue during the observation interval T = 10 min = 600 s.

Then, we use Eqs. (3.116) and (3.117) to calculate J and M

$$J = \tfrac{T}{2} \cdot (1 - \rho) - \tfrac{1}{C} \cdot (L_{s0} + 1) = \tfrac{600}{2} \cdot (1 - 1.125) - \tfrac{1}{0.72} \cdot (5 + 1) = -59.56\text{s}$$

$$M = \tfrac{4}{C} \cdot \left[\tfrac{T}{2} \cdot (1 - \rho) + \tfrac{1}{2} \cdot \rho \cdot T\right] = \tfrac{4}{0.272} \cdot \left[\tfrac{600}{2} \cdot (1 - 1.125) + \tfrac{1}{2} \cdot 1.125 \cdot 600\right]$$
$$= 4411.76\text{s}$$

Finally, from Eq. (3.115), we have

$$<w_s> = \frac{1}{2}\left(\sqrt{J^2 + M} - J\right) = \frac{1}{2}\left(\sqrt{(-59.56)^2 + 4411.76} + 59.56\right) = 74.39\ \mathrm{s}$$

Applying the same procedure that led to Eq. (3.115) for the transition relationship between the mean of the waiting time in the queue $E[w_c]$ under steady-state conditions (See the formulas shown in Table 3.3.)

$$E[w_c] = \frac{1}{C} \cdot \left(\frac{\rho}{1 - \rho}\right) \tag{3.118}$$

and the mean deterministic value $\overline{w}_c$

$$\overline{w}_c = \frac{L_{c0}}{C} + \frac{(\rho - 1) \cdot T}{2} \tag{3.45}$$

we obtain the time-dependent relationship $<w_c>$

$$<w_c> = \frac{1}{2}\left(\sqrt{P^2 + Q} - P\right) \tag{3.119}$$

with

$$P = \frac{1}{2} \cdot (1 - \rho) - \frac{1}{C} \cdot (L_{s0} - 1) \tag{3.120}$$

$$Q = \frac{2 \cdot T}{C} \cdot \left[\rho + \frac{2 \cdot L_{s0}}{C \cdot T}\right] \tag{3.121}$$

With the data of the worked example just developed, we evaluate the average waiting time in the queue $<w_c>$.

From Eqs. (3.120) and (3.121) we have

$$P = \tfrac{T}{2} \cdot (1 - \rho) - \tfrac{1}{C} \cdot (L_{s0} - 1) = \tfrac{600}{2} \cdot (1 - 1.125) - \tfrac{1}{0.272} \cdot (5 - 1) = -52.21\ \mathrm{s}$$

$$Q = \tfrac{2 \cdot T}{C} \cdot \left[\rho + \tfrac{2 \cdot L_{s0}}{C \cdot T}\right] = \tfrac{2 \cdot 600}{0.272} \cdot \left[1.125 + \tfrac{2 \cdot 5}{0.272 \cdot 600}\right] = 5233.56\ \mathrm{s}$$

and, therefore, with Eq. (3.119), for $<w_c>$

$$< w_c > = \frac{1}{2}\left(\sqrt{P^2 + Q} - P\right) = \frac{1}{2}\left(\sqrt{(-52.21)^2 + 5233.56} + 52.21\right) = 70.71\ \mathrm{s}$$

We note that

$$<w_s> - <w_c> = 74.39 - 70.71 = 3.68\ \mathrm{s}$$

This result is in accordance with Eq. (3.4) since $<w_s> - <w_c> = 3.68$ s, as indicated by Eq. (3.2), is the value of the mean of service time $E[T_s] = 1/C = 1/0.272 = 3.68$ s.

3.3.3 Generalized Time-Dependent Solutions

The previous relationships for the time-dependent mean values of the numbers of users in the system, the queue length and waiting times ($< L_S >$, $<L_C>$, $<w_S>$, and $<w_C>$) were obtained starting with the expressions of the expected values $E[L_S]$, $E[L_c]$, $E[w_S]$, and $E[w_c]$ under steady-state conditions that are valid in the case of Poissonian distributed leg arrivals and exponential service times (See Sect. 3.1).

The above-mentioned formulas for the process of arrivals and service times are generally considered to be sufficient to represent most of the situations of practical interest.

However, if one wants to have more general expressions than the ones already given by Eqs. (3.103) and (3.115), for example for $<L_S>$ and $<w_S>$, respectively, we can use Eqs. (3.13) and (3.11) to obtain the expected values $E[L_S]$ and $E[w_S]$ under steady-state conditions. From these equations, we can start the transition to the deterministic means $\overline{L}_S$ and $\overline{w}_S$.

$$E[L_S] = Q_e \cdot \overline{s} + \frac{Q_e^2 \cdot (\overline{s}^2 + V[s])}{2 \cdot (1 - Q_e \cdot \overline{s})} \tag{3.13}$$

$$E[w_S] = \overline{s} + \frac{Q_e \cdot (\overline{s}^2 + V[s])}{2 \cdot (1 - Q_e \cdot \overline{s})} \tag{3.11}$$

where the mean $\overline{s} = E[T_s]$ and the variance $V[s] = VAR[T_S]$ of the service time are generally given by Eqs. (3.8) and (3.9).

In practical applications, $\overline{s}$ is evaluated, as usual, as $\overline{s} = 1/C$, even though this choice may lead to distorted and mutually non-congruent estimations of $\overline{s}$ and $V[s]$, while, for $V[s]$, one may use the values obtained from traffic data that are available in the literature.

Combining Eq. (3.11) with Eq. (3.49) and using the same procedure illustrated in Sects. 3.3.1 and 3.3.2, we have, for example for $< w_S>$, the expression that generalizes Eq. (3.115) [8] using the simplifying assumption that $L_{S0} = 0$.

$$<w_S> = \overline{s} + (\overline{s} + V[s]) \cdot \frac{\rho \cdot C}{2} + \frac{T}{4}\left[\rho - 1 + \sqrt{(\rho - 1)^2 + \frac{4 \cdot \rho}{C \cdot T}}\right] \tag{3.122}$$

In Fig. 3.14, Eq. (3.122) is shown using these values: $\overline{s} = 6\,s$; $V[s] = 36$ s; $T = 10$ min $= 600$ s; $C = 540$ veh/h $= 0.15$ veh/s.

Starting with Eqs. (3.11) and (3.13), relationships similar to Eq. (3.122) can be obtained that generalize Eqs. (3.103), (3.107), and (3.119). Their deduction is left to the keen reader.

---

[8] The behaviors assumed for Q_e and C are again, as in the other cases analyzed so far, the ones shown in Fig. 3.8.

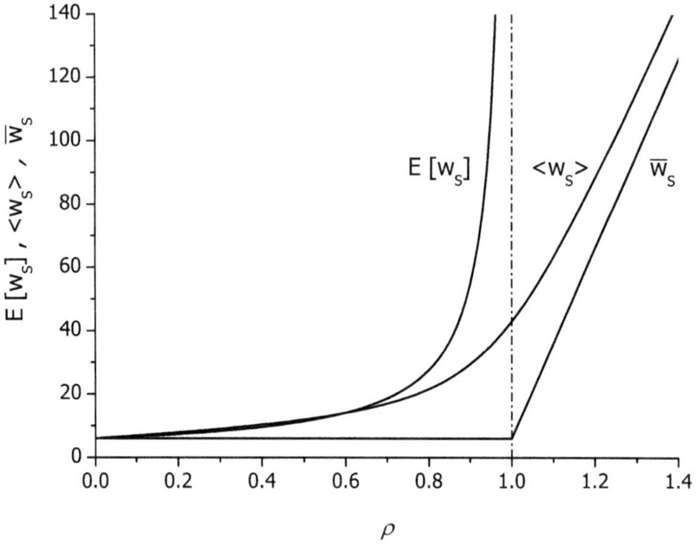

Fig. 3.14 Generalized relationship $<w_s> = < w_s(\rho) >$ for a case with $L_{s0} = 0$

3.3.4 *Time-Dependent Solutions for Traffic Peaks Between Two Periods at Steady-State Conditions*

We refer here to the behaviors of traffic demand and capacity shown in Fig. 3.15.

The conditions just before and just after the peak period are the same, and they both show undersaturation

$$Q_{e0} = Q_{e1} \qquad C_0 = C_1 \tag{3.123}$$

$$\rho_0 = \rho_1 < 1 \tag{3.124}$$

At the instant t_0, a traffic peak occurs, with $Q_e > Q_{e0}$ and $C < C_0$, and the entry is oversaturated ($\rho > 1$).

At the end of the peak period for which the duration is T, the entry instantly returns to undersaturation conditions at $t = t_0 + T$ (Eq. (3.123)).

It is straightforward to note that the behaviors of the cumulative arrival A(t) and departure D(t) values are of the type shown in Fig. 3.10 for this case when

$$L_{c0} = L_{c1} \tag{3.125}$$

For the reader's convenience, the above-mentioned behaviors are also shown in Fig. 3.15.

Through the transformation of coordinates technique (Eq. (3.92)) or (Eq. (3.95)), for the case being discussed here that starts with the same steady-state solution ($E[w_c]$, $E[w_s]$), it is possible to identify, in principle, more than one time-dependent

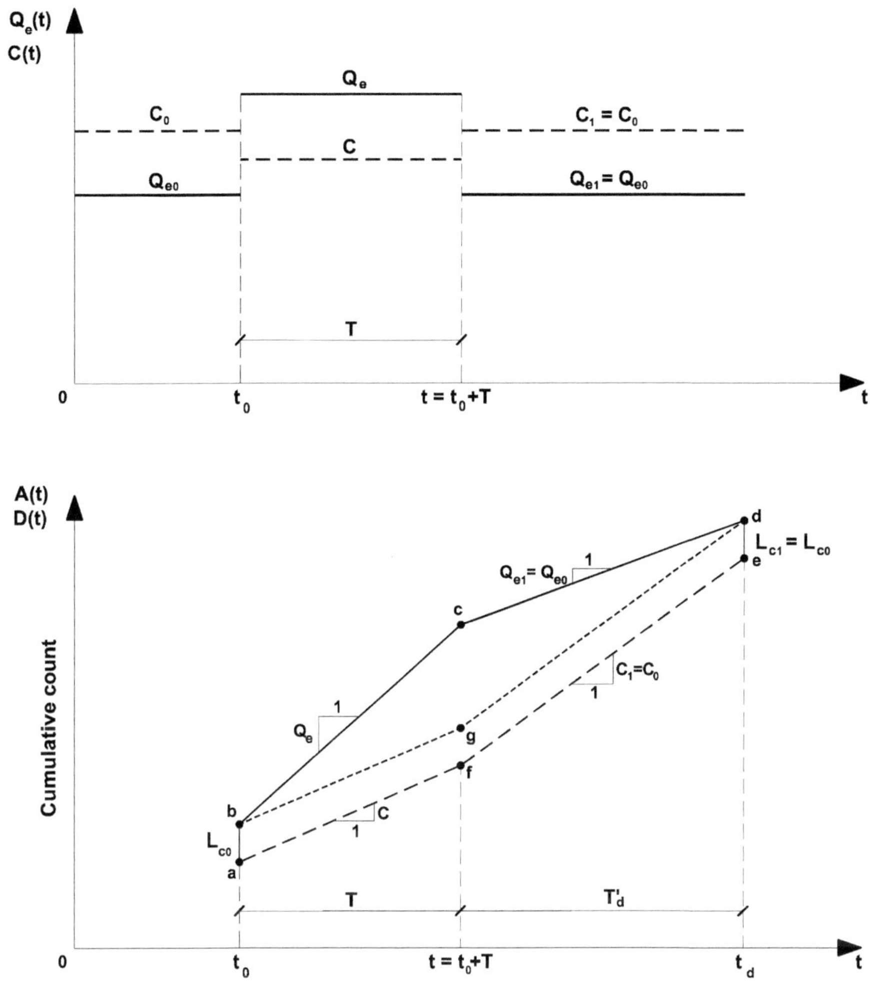

Fig. 3.15 Traffic peak between two steady-state conditions

relationship relative to the mean waiting times $<w_C>$ and $<w_S>$ under a transient condition, according to the deterministic formula on which the deductions are based.

In previous Sects. 3.2.2 and 3.2.3, we gave various relationships for $\overline{w}_C$ and $\overline{w}_S$ in the case of a queue at the beginning and at the end of the peak period T, each valid under specific assumptions about the number of vehicles on which we must distribute the total waiting time W_c caused by the traffic peak Q_e (only the vehicles that arrived during T or all those affected by the traffic peak).

However, by most of the above-mentioned relationships, it is not easy (or possible) to obtain closed-form relationships for $<w_C>$ and $<w_S>$.

Thus, among the time-dependent solutions obtainable, the most widely used today is still the one developed by Kimber and Hollis [4] that we will now derive.

We operate in terms of increase in the waiting times with respect to the steady-state conditions.

If the system instantly passes from a steady-state condition (characterized by a traffic intensity value ρ_0) to another condition (characterized by a traffic intensity value ρ), the difference $\Delta E[w_c]$ between the corresponding expected values of the waiting times $E[w_{c0}]$ and $E[w_c]$, if the arrivals are Poissonian and the service times are exponential (See Table 3.3) is

$$\Delta E[w_c] = \frac{1}{C} \cdot \left(\frac{\rho}{1-\rho}\right) - \frac{1}{C_0} \cdot \left(\frac{\rho_0}{1-\rho_0}\right) \tag{3.126}$$

The deterministic increase $\Delta \overline{w}_c$ of the average waiting time in the queue with respect to $E[w_c]$, caused by the temporary oversaturation of the entry during T, may be evaluated. This is done by starting with Eq. (3.86) and subtracting from the area $w_c^{**} = S(abcdefa)$, which gives the total waiting time in the queue (see Fig. 3.15), the area S(abgdefa), which has the same value of the cumulative probabilistic waiting time. Thus, we have, with B from Eq. (3.56):

$$\Delta \overline{w}_c = \frac{1/2 \cdot (A^{**} + B) - L_{c0}(T + T'_d)}{Q_e \cdot T} = \frac{1/2 \cdot (A^{**} + B) - L_{c0}(T + T'_d)}{\rho \cdot C \cdot T} \tag{3.127}$$

where A^{**} must be evaluated by substituting $L_{c1}^2 = L_{c0}^2$ into Eq. (3.68).

In terms of traffic intensity ρ and capacity C before, during, and after the peak, Eq. (3.127), by means of simple transformations, becomes

$$\Delta \overline{w}_c = \frac{1}{2} \cdot \left(\frac{\rho - 1}{\rho}\right) \cdot \left[\frac{(1-\rho_0) \cdot C_0 + (\rho - 1) \cdot C}{(1-\rho_0) \cdot C_0}\right] \cdot T \tag{3.128}$$

To obtain the transition curve $\Delta <w_c>$ from $\Delta E[w_c]$ to $\Delta \overline{w}_c$, we use Eq. (3.92) or Eq. (3.95).

$$\rho_e = \rho_T - (\rho_d - 1) \tag{3.95}$$

Even in this case (See Fig. 3.12), we indicate the generic values of traffic intensity with ρ_e, ρ_T, and ρ_d, with the following results

$$\Delta E[w_c] = \Delta \overline{w}_c = \Delta <w_c> \tag{3.129}$$

Now, to make ρ_d from Eq. (3.128) explicit after it has been substituted for ρ, we use transformations of variables to simplify the calculation

$$(1 - \rho_0) \cdot C_0 = C - y \cdot \rho \tag{3.130}$$

Thus, using Eqs. (3.130) and (3.129), we obtain

$$\rho_d = \frac{C \cdot (2 \cdot \Delta \overline{w}_c/T) + C - y}{y \cdot (2 \cdot \Delta \overline{w}_c/T) + C - y} \tag{3.131}$$

that is, since, by Eq. (3.129), in correspondence of ρ_d, $\Delta \overline{w}_c = \Delta <w_c>$

$$\rho_d = \frac{C \cdot (2 \cdot \Delta <w_c>/T) + C - y}{y \cdot (2 \cdot \Delta <w_c>/T) + C - y} \tag{3.132}$$

With this last expression of ρ_d inserted into Eq. (3.95) we have

$$\rho_e = \rho_T - \left(\frac{C \cdot (2 \cdot \Delta <w_c>/T) + C - y}{y \cdot (2 \cdot \Delta <w_c>/T) + C - y} - 1 \right) \tag{3.133}$$

Now, we specify Eq. (3.126) with $\rho = \rho_e$, taking into account Eq. (3.129)

$$\Delta <w_c> = \frac{1}{C} \cdot \left(\frac{\rho_e}{1 - \rho_e} \right) - \frac{1}{C_0} \cdot \left(\frac{\rho_0}{1 - \rho_0} \right) \tag{3.134}$$

If we express ρ_e in Eq. (3.134) using its mathematical relationship from Eq. (3.133), we finally have the implicit equation of the transition curve $\Delta <w_c>$ between $\Delta E[w_c]$ and $\Delta \overline{w}_c$.

$$\Delta <w_c> + \frac{1}{C_0} \cdot \left(\frac{\rho_0}{1 - \rho_0} \right) - \left\{ \frac{\frac{1}{C} \cdot \left[\rho - \frac{\Delta <w_c>}{\left[\left(\frac{y}{C-y} \right) \cdot \Delta <w_c> + \frac{T}{2} \right]} \right]}{1 - \rho + \frac{\Delta <w_c>}{\left[\left(\frac{y}{C-y} \right) \cdot \Delta <w_c> + \frac{T}{2} \right]}} \right\} = 0 \tag{3.135}$$

Equation (3.135), solved with respect to $\Delta <w_c>$ with reference to only the first quadrant of the plane (ρ; increases of $\Delta(\cdot)$), since ρ_T is a current value of traffic intensity ($\rho_T = \rho$), leads to

$$\Delta <w_c> = \frac{1}{2} \left(\sqrt{A^2 + B} - A \right) \tag{3.136}$$

with

$$A = \frac{1/2 \cdot (1 - \rho) \cdot (C - y) \cdot T + 1/C \cdot [C - y \cdot (1 + \rho)]}{C - y \cdot \rho} + \frac{1}{C_0} \cdot \left(\frac{\rho_0}{1 - \rho_0} \right) \tag{3.137}$$

$$B = \frac{2 \cdot T \cdot \left[\frac{1}{C} \cdot \rho - (1 - \rho) \cdot \frac{1}{C_0} \cdot \frac{\rho_0}{(1 - \rho_0)} \right] \cdot (C - y)}{C - y \cdot \rho} \tag{3.138}$$

Finally, the average waiting time in the queue $<w_c>$ is obtained by adding the mean waiting time $E[w_c]$ at statistical equilibrium to the increase $\Delta<w_c>$

$$<w_c> = \Delta<w_c> + \frac{1}{C_0} \cdot \left(\frac{\rho_0}{1-\rho_0}\right) \tag{3.139}$$

From Eqs. (3.139) and (3.4), we obtain the average time spent in the system

$$<w_s> = \Delta<w_c> + \frac{1}{C_0} \cdot \left(\frac{\rho_0}{1-\rho_0}\right) + \overline{T}_s \tag{3.140}$$

Equation (3.139), deduced according to Kimber and Hollis, is generally regarded as the conventional average waiting time in the queue (See Sect. 3.2.3).

It was obtained starting with a total incremental delay[9] represented by the numerator of Eq. (3.127) that, although relative to all the vehicles affected by the traffic peak (vehicles that arrived during the peak period T and the following period T'_d, which is still affected by the peak effects), was distributed only over the vehicles that arrived during the peak interval (the denominator of Eq. (3.127)).

In other words, the peak vehicles that $<w_c>$ (Eq. (3.139)) refers to, are given a average waiting time in the queue greater than the one that really pertains to them. In fact, the total waiting time (Eq. (3.127)) includes the waits caused but not endured by peak vehicles. The waiting times that are not endured by the peak vehicles are instead endured by the vehicles arrived during T'_d, which are still affected by the effects of peak demand.

It follows that, for the mean $\overline{T}_s$ of the service time to insert into Eq. (3.140) to obtain from $<w_c>$, the mean $<w_s>$ of the times spent in the system should be use the weighted mean of the service times $1/C$ and $1/C_1 = 1/C_0$ on the respective times of application. (See illustrations in previous Sects. 3.2.1 and 3.2.2 concerning the calculation of $\overline{T}_s$.)

A simpler way would be to use only the mean value $\overline{T}_s = 1/C$ relative to the peak period T [8].

Besides, we may rewrite A (Eq. (3.137)) and B (Eq. (3.138)) in terms of demand (Q_e and $Q_{e0} = Q_{e1}$) and capacity (C and $C_0 = C_1$). Since $\rho = Q_e/C$ and $\rho_0 = \rho_1 = Q_{e0}/C_0$, y (Eq. (3.130)) is equal to $y = [C-(1-Q_{e0}/C_0)\cdot C_0]/(Q_e/C)$. From the Eq. (3.135), we obtain (flows and capacity veh/h; T in h):

$$<w_c> = \left[\frac{1}{2} \cdot \left(\sqrt{F^2 + G} - F\right) + E\right] \cdot 3600\ (s) \tag{3.141}$$

$$<w_s> = \left[\frac{1}{2} \cdot \left(\sqrt{F^2 + G} - F\right) + E + \frac{1}{C}\right] \cdot 3600\ (s) \tag{3.142}$$

9 with respect to the cumulative delay derived from the steady-state condition characterized by $\rho_0 < 1$.

with

$$F = \frac{1}{C_0 - Q_{e0}} \cdot \left[\frac{T}{2} \cdot (C - Q_e) \cdot z + \alpha \cdot \left(z - \frac{h}{C} \right) \right] + E \tag{3.143}$$

$$G = \frac{2 \cdot T \cdot z}{C_0 - Q_{e0}} \cdot \left[\alpha \cdot \frac{Q_e}{C} - (C - Q_e) \cdot E \right] \tag{3.144}$$

$$E = \frac{\alpha \cdot Q_{e0}}{C_0 \cdot (C_0 - Q_{e0})} \tag{3.145}$$

$$h = C - C_0 + Q_{e0} \tag{3.146}$$

$$z = 1 - \frac{h}{Q_e} \tag{3.147}$$

The parameter α was introduced by Kimber and Hollis in addition to the variables of the starting formula (Eq. (3.136)). It was added after the validation (and thus setting) of Eq. (3.136) in order to make a comparison with the results obtained by a different calculation criterion for waiting phenomena under a transient condition.

The parameter α was introduced to take into account peak periods T of very short duration (e.g., T = 2–3 min) and the suggested value for α is 2.

With $\alpha = 2$, however, one obtains unrealistic results for values of peak periods that occur more frequently (e.g., with durations in the range of 10–15 min), so, under these circumstances, one must use $\alpha = 1$.

Regarding the calculation of the percentiles $L_{s,p}$ of the number of users in the system L_s during the traffic peak, it implies the determination of the distribution functions $F_{Ls}(\cdot)$ of L_s under a transient state.

Also, for $F_{Ls}(\cdot)$ as was just illustrated for $<w_c>$ and $<w_s>$, we can heuristically use the transformation of coordinates technique (Eq. (3.95)), because it is difficult to use the exact results of the mathematical queuing theory under a transient condition.

With the above-mentioned criterion applied to the transition, Wu [9] obtained $F_{Ls}(\cdot)$ and thus the expression of the percentiles $L_{s,p}$ of the number of users in the system during the traffic peak. To do so, Wu used the formulation of the distribution function $F_{Ls}(\cdot)$ at a statistical equilibrium of L_s to the deterministic law of the same parameter as a function of traffic intensity ρ.

The deductive procedure of $F_{Ls}(\cdot)$ and, thus, of the percentiles $L_{s,p}$, is shown in a detailed and clear way in [10]. We refer to it for the development of the calculations that lead to the following relationship

$$L_{s,p} = \frac{C \cdot T}{4} \left[\rho - 1 + \sqrt{(1 - \rho)^2 + \frac{8\rho}{C \cdot T} \left[-\ln \gamma \right]} \right] \tag{3.148}$$

where

$$\gamma = \left(1 - \frac{p}{100} \right) \tag{3.149}$$

Thus, for example, the estimation of the 95th percentile of L_s (with $\gamma = 0.05$) is

$$L_{s,95} = \frac{C \cdot T}{4} \left[\rho - 1 + \sqrt{(1-\rho)^2 + \frac{8\rho}{C \cdot T} \cdot 3} \right] \tag{3.150}$$

while, with $\gamma = 0.01$, the 99th percentile of L_s is obtained from the relationship

$$L_{s,99} = \frac{C \cdot T}{4} \left[\rho - 1 + \sqrt{(1-\rho)^2 + \frac{8\rho}{C \cdot T} \cdot 4{,}6} \right] \tag{3.151}$$

However, in steady-state conditions, as a first approximation, the double of the average value of $L_{s,p}$ generally represents an acceptable estimation for the application of a percentile p sufficiently high (p = 90–95) of the number of users in the system, i.e. $L_{s,p} \cong 2\,E[L_s]$.

In the next chapter, we will illustrate some detailed practical applications of the time-dependent solutions presented in this section.

3.3.5 Concluding Remarks

Except for Eq. (3.122), the time-dependent relationships described in the previous sections were obtained by assuming that statistical equilibrium solutions, which then allow us to reach deterministic solutions, are relative to Poissonian arrivals and exponential service times. Also, in the case illustrated in Sect. 3.3.4, the assumption was made that statistical equilibrium conditions before and after the peak period are the same ($Q_{e0} = Q_{e1}$; $C_0 = C_1$).

The plausibility of these assumptions is proved by theoretical research and by the good results in technical practice.

More general relationships than the ones deduced starting with the formulation just recalled of Poissonian arrivals and exponential service times (for example, Eq. (3.122)) are described in [4], and the reader should refer to this reference for additional information.

However, regardless of the levels of generalization used, the results obtained by the time-dependent formulas, given their heuristic nature, do not completely agree with the exact results obtained through the mathematical queuing theory in the absence of statistical equilibrium.

However, the discrepancies for the practical applications of interest are negligible.

As we have often stressed in this chapter, time-dependent relationships (See Fig. 3.12) are continuous functions of traffic intensity ρ. This ρ can vary inside its entire domain interval ($\rho \in [0;+\infty[$).

This allows us to obtain solutions in the neighborhood of $\rho = 1$ (where both probabilistic and deterministic approaches give unrealistic results) and to treat the case of a traffic peak that is characterized by entry undersaturation during the peak period.

In other words, it is possible to examine demand situations that can be defined as peak situations, since the value of Q_e in T is greater than Q_{e0} and Q_{e1}, which pertain to the intervals before and after T ($Q_e > Q_{e0}$; $Q_e > Q_{e1}$), respectively, but such that the entry remains constantly undersaturated ($Q_e < C$, i.e $\rho < 1$) for each t.

This type of circumstance will be examined in detail through some worked examples in the next chapter.

Finally, it is worth noting that the approximate solutions for studying the transient states of waiting phenomena at intersections are being thoroughly investigated [10]. This proves that this topic is considered to be essential in the field of functional design of intersections.

References

1. Cassandras, C. G., *Introduction to Discrete Event Systems*, Springer, New York, 1999.
2. Louah, G., Priority Intersections: Modelling, in *Concise Encyclopedia of Traffic and Transportation System* (Papageorgiu, M., Editor), Pergamon Press, Oxford, 1991.
3. Kleinlock, L., *Queuing Systems – Vol. I*, John Wiley and Sons, New York, 1975.
4. Kimber, R. M., Hollis, E. M., *Traffic Queues and Delays at Road Junctions*, TRRL Laboratory Report 909, Crowthorne, Berkshire, 1979.
5. Brilon, W., *Delays at Oversaturated Unsignalized Intersections, Based on Reserve Capacities*, Transportation Research Record 1484, Washington DC, 1995.
6. Newell, G. F., *Queues – Inventories and Maintenance*, 3rd edition, John Wiley & Sons, New York, 1976.
7. Morse, P. M., *Application of Queuing Theory*, 2nd edition, Chapman Hall, London, 1982.
8. Kimber, R. M., Hollis, E. M., *Peak-Period Traffic Delays at Road Junctions and Other Bottlenecks*, Traffic Engineering and Control (TEC), October 1978.
9. Wu, N., *An Approximation for Distribution of Queue Lengths at Unsignalized Intersections*, Proceedings of Second International Symposium on Highway Capacity, Vol. 2, ARRB, Sydney, 1994.
10. Brilon, W., *Time Dependent Delay at Unsignalized Intersections*, Proceedings of 17th International Symposium on Transportation and Traffic Theory (ISTTT17), London, July 2007.

Chapter 4
Calculation of Waiting Times, Queue Lengths, and Levels of Service

With the formulas described in the previous chapter, it is possible to evaluate the waiting times and the number of vehicles in the queue and in the system. This can be done by following the time evolution of traffic demand at legs and by the calculation procedure of capacity given in Chap. 2.

According to the level of description of demand variation in time, different computational procedures are used.

If demand Q_e is considered for successive limited[1] period of T_k (generally having the same Δt value) and is approximated by a continuous function (See Fig. 4.1), the state variables may be performed following step by step the system evolution (See Sect. 4.2). If Q_e values are relative to greater[2] time periods (or by adding determinations relative to successive periods, See Fig. 4.2), the state variables can be obtained by the procedure of Sect. 3.3.4.

The choice among the above-mentioned options depends on various considerations, among which are the availability of more or less detailed measurements (or estimations) of Q_e and the aims of the analysis.

In the remaining part of this chapter, we will illustrate some important practical applications of the results described in Chap. 3 to roundabouts, starting with the case of a system evolution in which the entries are systematically undersaturated.

4.1 State Evolutions with Undersaturated Entries

First, let us consider the case of system evolution among conditions that can be considered as steady-state.

We recall that, for technical aims, a steady-state condition is reached for undersaturated entries, if the traffic demand at the intersection is constant for a finite time period T, which is long enough to allow the stabilization of the operating conditions

---

[1] In current technical practice, the values of these intervals are generally in the range of 5–15 min.

[2] In current technical practice, the values of these intervals are generally in the range of 30 min to an hour.

R. Mauro, *Calculation of Roundabouts*, DOI 10.1007/978-3-642-04551-6_4,

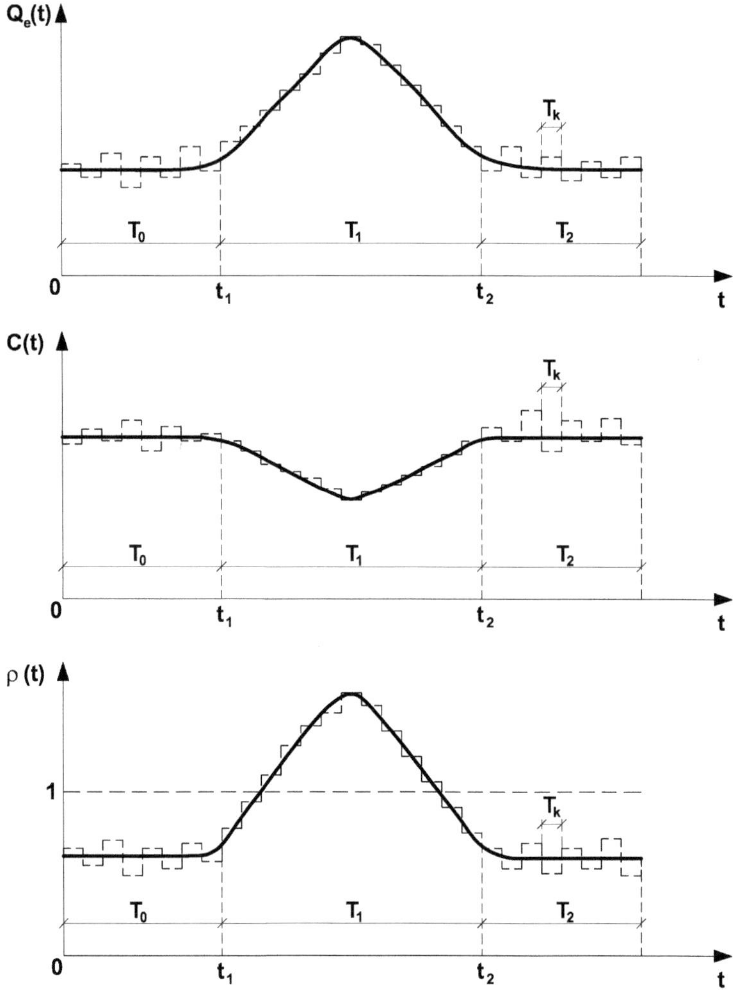

Fig. 4.1 Time evolution of traffic demand, capacity, and traffic intensity at an entry approximated by continuous functions

of the roundabout in the neighborhood of constant mean values of the state variables. In addition, the punctual values of state variables must be little dispersed around the mean values E [·].

The transition from one to another of these statistical equilibrium conditions is assumed as instantaneous.

In this connection, it is possible to demonstrate [1] that an approximated measurement of the time ϑ_i, which the system must pass, at a generic entry "i", from a steady condition, characterized by a traffic intensity ρ_{i0}, to another steady condition characterized by ρ_{i1}, is given by (footnote no. 4 in Sect. 1.1)

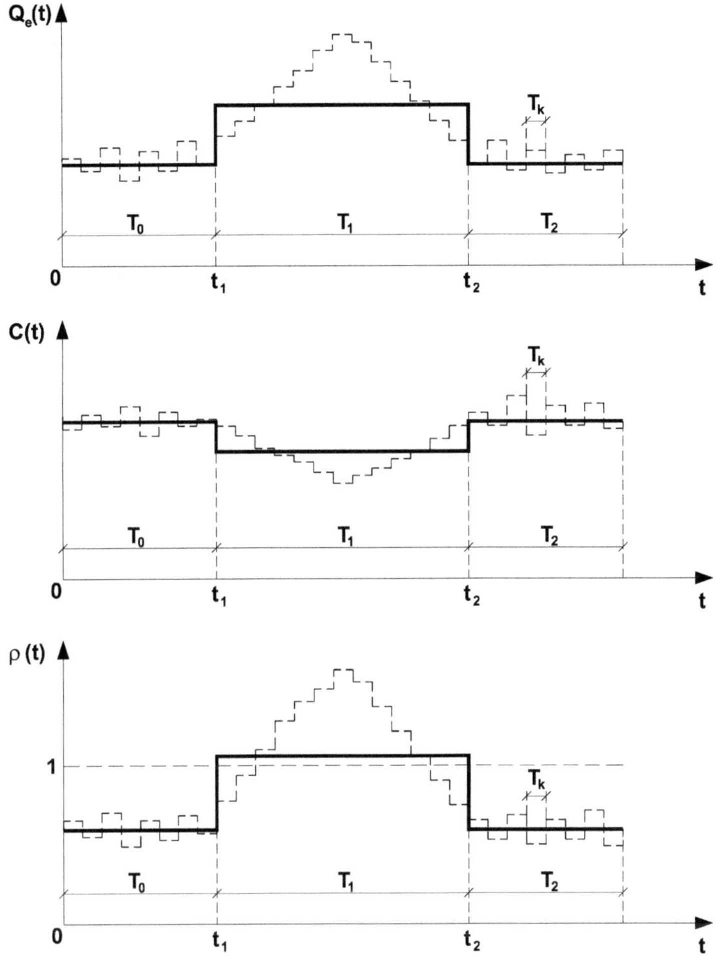

Fig. 4.2 Time evolution of traffic demand, capacity, and traffic intensity at an entry approximated by step functions

$$\vartheta_i = \frac{1}{(\sqrt{C_i} - \sqrt{Q_i})^2} = \frac{1}{C_i \cdot (1 - \sqrt{\rho_{i1}})^2} \qquad (\rho_{i1} \leq 1) \tag{4.1}$$

If $\rho_{i1} \ll 1$, the value of ϑ_i is small, and the new statistical equilibrium condition is reached rapidly (for practical purposes, instantaneously). If ρ_{i1} tends to 1, ϑ_i becomes large, and, in this case, transient state conditions must be taken into account during ϑ_i.

In other words, if demand (and/or capacity) varies with respect to the previous state for a period T_i, it is possible to neglect the time-dependent aspects of the waiting phenomenon if

$$\rho_{i1} < \rho_{ic} \tag{4.2}$$

with

$$\rho_{ic} = \left(1 - \frac{1}{\sqrt{C_i \cdot T_i}}\right)^2 \tag{4.3}$$

If, instead, $\rho_{i1} > \rho_{ic}$, in the analysis of the system, the results of the statistical equilibrium are not usable, but time-dependent formulas are used.

When the roundabout evolves through conditions that can be considered as steady-state conditions, the determinations of the state variables as a function of capacity C_i and of traffic intensity $\rho_i = Q_{ei}/C_i$ are performed with the formulas shown in Table 3.3 for Poissonian arrivals and exponential service times.

It is worth noting that the entering demand flows must be increased by the number of vehicles waiting at the end of the previous time segment T_{k-1} for the calculation of capacity by assuming that capacity is constant at intervals in the various successive intervals T_k of subdivision of T.

In other words, if the interval T_k demand at the generic leg "i" is equal to $Q_{ei}^{(k)}$ and several vehicles $L_{si}^{(k-1)}$ are already at the entry, the flow entering the circulatory roadway during T_k, $Q_{ei}^{(k)*}$, with $Q_{ei}^{(k)}$ and $Q_{ei}^{(k)*}$ expressed in pcu/h, is equal to

$$Q_{ei}^{(k)*} = \left(Q_{ei}^{(k)} \cdot \frac{T_k}{60} + L_{si}^{(k-1)}\right) \cdot \frac{60}{T_k} = Q_{ei}^{(k)} + L_{si}^{(k-1)} \cdot \frac{60}{T_k} \tag{4.4}$$

where the number of users in the system $L_{si}^{(k-1)}$ is [3]

$$L_{si}^{(k-1)} = E[L_{si}]^{(k-1)} \text{ at steady-state conditions} \tag{4.5}$$

$$L_{si}^{(k-1)} = \langle L_{si} \rangle^{(k-1)} \text{ at transient conditions} \tag{4.6}$$

If we record zero values of traffic demand at a generic entry "i" in T_k, with Eq. (4.4) it is possible to take into account a flow entering the circulatory roadway that consists of $L_{si}^{(k-1)}$ vehicles in the system that queued up during the interval T_{k-1}.

From Eq. (4.4), it is straightforward to deduce that, if the interval T_k is sufficiently long, the contribution of the waiting vehicles $L_{si}^{(k-1)}$ to the entering flow is negligible, and it becomes null when T is infinite.

Now, it is worth noting that, for calculation aims, in all the relations relative to the waiting phenomenon described in Chap. 3, traffic demand and capacity must be expressed in vehicles (veh) per time, generally veh/h or veh/s.

---

[3] We recall that, with $E[\cdot]$, we indicate the expected value and with $\langle \cdot \rangle$, we indicate the time-dependent mean (Sect. 3.3).

In the capacity formulas (See Chap. 2), demand and capacity are measured in passenger car units (pcu) per time, i.e., generally as pcu/h or pcu/s.

Therefore, the values of traffic demand and the values obtained by capacity calculations must be converted from one type of measurement to the other in order to be used to determine waiting times, queue length, and the number of vehicles in the system.

Capacity and traffic demand values Q_e and C (in veh/time units) are obtained by multiplying the determinations of Q_e' and C' by the self-evident factor f if Q_e' and C' are given in pcu/time units

$$f = \frac{1}{(1 - P_p - P_{cm}) + \alpha_p \cdot P_p + \alpha_{cm} \cdot P_{cm}} \tag{4.7}$$

where P_p and P_{cm} are the rates of heavy vehicles (coaches included) and bicycles and motorbikes, respectively, present in the flow Q_e. The symbols α_p and α_{cm} are used to indicate the coefficients of equivalence of the above-mentioned vehicles in passenger cars (in general, see Sect. 2.1, $\alpha_p = 2$ and $\alpha_{cm} = 0.5$).

On the other hand, the determination of $L_{si}^{(\cdot)}$ that is expressed in vehicle units (veh), in order to be added in Eq. (4.4) to the quantity $Q_{ei}^{(\cdot)} \cdot T./60$ (where $Q_{ei}^{(\cdot)}$ is in pcu/h) must be converted, using the factor f provided by Eq. (4.7), into passenger car units (pcu).

To calculate traffic intensity ρ, we use one or the other vehicular volume measure, thus obtaining the same numerical result.

To better explain the above-mentioned points, we will give some calculation examples in the following sections.

4.1.1 First Worked Example

Consider a four-legged roundabout with a single-lane circulatory roadway and single-lane entries.

Traffic demand (given in the form shown in Sect. 1.1.1) evolves instantaneously through three stages.

To simplify the procedure, we assume that the percentages of vehicles different from passenger cars are completely negligible, so that the values of demand and capacity expressed in pcu/h coincide with those expressed in veh/h.

State 0 (from $t = 0$ to $t = t_1$)

$$[Q_{ei}^{(0)}] = \begin{bmatrix} 680 & 600 & 731 & 550 \end{bmatrix} \; P_{O/D}^{(0)} \equiv \begin{bmatrix} 0 & 0.40 & 0.40 & 0.20 \\ 0.35 & 0 & 0.50 & 0.15 \\ 0.15 & 0.30 & 0 & 0.55 \\ 0.40 & 0.40 & 0.20 & 0 \end{bmatrix} \tag{4.8}$$

$$M_{O/D}^{(0)} \equiv \begin{bmatrix} 0 & 272 & 272 & 136 \\ 210 & 0 & 300 & 90 \\ 110 & 219 & 0 & 402 \\ 220 & 220 & 110 & 0 \end{bmatrix} \tag{4.9}$$

State 1 (from $t = t_1$ to $t = t_2$ for a duration $T_1 = 20$ min)

$$[Q_{ei}^{(1)}] = \begin{bmatrix} 350 & 280 & 404 & 309 \end{bmatrix} \; P_{O/D}^{(1)} \equiv \begin{bmatrix} 0 & 0.25 & 0.36 & 0.39 \\ 0.29 & 0 & 0.37 & 0.34 \\ 0.33 & 0.29 & 0 & 0.38 \\ 0.31 & 0.35 & 0.34 & 0 \end{bmatrix} \tag{4.10}$$

$$M_{O/D}^{(1)} \equiv \begin{bmatrix} 0 & 88 & 126 & 136 \\ 81 & 0 & 104 & 95 \\ 132 & 116 & 0 & 156 \\ 96 & 108 & 105 & 0 \end{bmatrix} \tag{4.11}$$

State 2 (from $t = t_2$ for a duration $T_2 = 30$ min)

$$[Q_{ei}^{(2)}] = \begin{bmatrix} 455 & 460 & 433 & 420 \end{bmatrix} \; P_{O/D}^{(2)} \equiv \begin{bmatrix} 0 & 0{,}35 & 0{,}35 & 0{,}30 \\ 0{,}30 & 0 & 0{,}35 & 0{,}35 \\ 0{,}15 & 0{,}30 & 0 & 0{,}55 \\ 0{,}20 & 0{,}40 & 0{,}40 & 0 \end{bmatrix} \tag{4.12}$$

$$M_{O/D}^{(2)} \equiv \begin{bmatrix} 0 & 159 & 159 & 137 \\ 138 & 0 & 161 & 161 \\ 65 & 130 & 0 & 238 \\ 84 & 168 & 168 & 0 \end{bmatrix} \tag{4.13}$$

To calculate capacity, we use the German formula by Brilon-Wu (Eq. 2.12).

State 0

By applying Eq. (1.12) from Chap. 1 to the elements of matrix (4.9), we obtain the circulating flows $[Q_{ci}^{(0)}]$

$$[Q_{ci}^{(0)}] = [\, Q_{c1}^{(0)} \; Q_{c2}^{(0)} \; Q_{c3}^{(0)} \; Q_{c4}^{(0)} \,] = [\, 549 \; 518 \; 436 \; 539 \,] \, (\text{pcu/h}) \tag{4.14}$$

and, similarly, the capacities are obtained by applying Eq. (2.12) of Chap. 2

$$[C_i^{(0)}] = [\, C_1^{(0)} \; C_2^{(0)} \; C_3^{(0)} \; C_4^{(0)} \,] = [\, 776 \; 800 \; 866 \; 784 \,] \, (\text{pcu/h}) \tag{4.15}$$

For the traffic intensities (degrees of saturation) $\rho_i{}^{(0)} = Q_{ei}{}^{(0)}/C_i{}^{(0)}$, we have

$$[\rho_i^{(0)}] = [\rho_1^{(0)} \; \rho_2^{(0)} \; \rho_3^{(0)} \; \rho_4^{(0)} \,] = [\, 0.88 \; 0.75 \; 0.84 \; 0.70 \,] \tag{4.16}$$

For each entry lane, the capacities and traffic intensities are known, so, using the relationships shown in Table 3.3 of Chap. 3, we calculate the averages of the times spent in the system $E[w_{si}]^{(0)}$ and the number of the vehicles in the system $E[L_{si}]^{(0)}$

$$[E[w_{si}]^{(0)}] = [E[w_{s1}]^{(0)}\ E[w_{s2}]^{(0)}\ E[w_{s3}]^{(0)}\ E[w_{s4}]^{(0)}] = \\ = [\,38.66\ 17.99\ 25.98\ 15.31\,]\ (s) \tag{4.17}$$

$$[E[L_{si}]^{(0)}] = [E[L_{s1}]^{(0)}\ E[L_{s2}]^{(0)}\ E[L_{s3}]^{(0)}\ E[L_{s4}]^{(0)}] = \\ = [\,7.33\ 3.00\ 5.25\ 2.33\,]\ (veh) \tag{4.18}$$

For the vector of the percentiles, we have $[L_{si,p}]^{(0)} \cong 2\ [E[L_{si}]^{(0)}]$, (p = 90–95, see the end of Sect. 3.3.4) and, therefore

$$[L_{si,p}]^{(0)} = [\,14.66\ 6.00\ 10.50\ 4.66\,]\ (veh) \tag{4.19}$$

State 1

To calculate the circulating flows $[Q_{ci}^{(1)}]$, we take into account the number of vehicles $[E[L_{si}]^{(0)}]$ waiting at legs calculated for the previous condition (vector (4.18)) that enter the circulatory roadway during the interval $T_1 = 20$ min and are therefore added to those that form demand flow $[Q_{ei}^{(1)}]$.

In conclusion, $[Q_{ei}^{(1)}]$ (Eq. (4.10)) and $[E[L_{si}]^{(0)}]$ (Eq. (4.18)) yield (Eq. (4.4)) the volumes (expressed in pcu/h)[4]

$$\begin{aligned} Q_{e1}^{(1)*} &= 350 + 7.33 \cdot \tfrac{60}{20} = 372 \\ Q_{e2}^{(1)*} &= 280 + 3.00 \cdot \tfrac{60}{20} = 289 \\ Q_{e3}^{(1)*} &= 404 + 5.25 \cdot \tfrac{60}{20} = 420 \\ Q_{e4}^{(1)*} &= 309 + 2.33 \cdot \tfrac{60}{20} = 316 \end{aligned} \tag{4.20}$$

that is

$$[Q_{ei}^{(1)*}] = \left[\,372\ 289\ 420\ 316\,\right]\ (pcu/h) \tag{4.21}$$

With the elements of vector (4.21) and matrix (4.10), Eq. (1.12) of Chap. 1 yields the following circle flows

$$[Q_{ci}^{(1)*}] = \left[\,340\ 386\ 327\ 344\,\right]\ (pcu/h) \tag{4.22}$$

With these values, we have the determinations of entry capacities during $T_1 =$ 20 min from Eq. (2.12) of Chap. 2

[4] In this case, we recall that traffic demand consists of only passenger cars, and, therefore, it is not necessary to convert the number of waiting vehicles into passenger car units.

$$[C_i^{(1)}] = \left[945\ 909\ 956\ 942\right]\ \text{(pcu/h)} \tag{4.23}$$

With vectors (4.10) and (4.23), we calculate the traffic intensity vector $\rho_i{}^{(1)} = Q_{ei}{}^{(1)}/C_i{}^{(1)}$

$$[\rho_i^{(1)}] = [\rho_1^{(1)}\ \rho_2^{(1)}\ \rho_3^{(1)}\ \rho_4^{(1)}] = [\,0.37\ 0.31\ 0.42\ 0.33\,] \tag{4.24}$$

From Eq. (4.1), we obtain the values of each time interval $\vartheta_i{}^{(1)}$ using the capacities and traffic intensities described in Eqs. (4.23) and (4.24), respectively.

$$\begin{aligned}
\vartheta_1^{(1)} &= \frac{1}{C_1^{(1)}\cdot(1-\sqrt{\rho_1^{(1)}})^2} = \frac{1}{(945/3600)\cdot(1-\sqrt{0.37})^2} = 24.8\ \text{s}\\
\vartheta_2^{(1)} &= \frac{1}{C_2^{(1)}\cdot(1-\sqrt{\rho_2^{(1)}})^2} = \frac{1}{(907/3600)\cdot(1-\sqrt{0.31})^2} = 20.2\ \text{s}\\
\vartheta_3^{(1)} &= \frac{1}{C_3^{(1)}\cdot(1-\sqrt{\rho_3^{(1)}})^2} = \frac{1}{(956/3600)\cdot(1-\sqrt{0.42})^2} = 30.4\ \text{s}\\
\vartheta_4^{(1)} &= \frac{1}{C_4^{(1)}\cdot(1-\sqrt{\rho_4^{(1)}})^2} = \frac{1}{(942/3600)\cdot(1-\sqrt{0.33})^2} = 21.1\ \text{s}
\end{aligned} \tag{4.25}$$

All the values of $\vartheta_i^{(1)}$ obtained are much smaller than the observation interval $T_1 = 20$ min $= 1200$ s.

The assumption of instantaneous transition from stage 0 to stage 1 that we made at the beginning of this worked example is, therefore, plausible.

In other words, the interval $T_1 = 20$ min is sufficiently wide to allow the roundabout to rapidly reach steady-state conditions and preserve them in stage 1.

The averages of the times spent in the system (in s) $E[w_{si}]^{(1)}$ and the number of vehicles in the system $E[L_{si}]^{(1)}$ are obtained by using the relationships shown in Table 3.3 of Chap. 3 (Poissonian arrivals and exponential service times) and the values of C_i and ρ_i described in Eqs. (4.23) and (4.24), respectively:

$$\begin{aligned}
[E[w_{si}]^{(1)}] &= [E[w_{s1}]^{(1)}\ E[w_{s2}]^{(1)}\ E[w_{s3}]^{(1)}\ E[w_{s4}]^{(1)}\,] =\\
&= [\,6.05\ 5.75\ 6.50\ 5.71\,]\ \text{(s)}
\end{aligned} \tag{4.26}$$

$$\begin{aligned}
[E[L_{si}]^{(1)}] &= [E[L_{s1}]^{(1)}\ E[L_{s2}]^{(1)}\ E[L_{s3}]^{(1)}\ E[L_{s4}]^{(1)}\,] =\\
&= [\,0.59\ 0.45\ 0.72\ 0.49\,]\ \text{(veh)}
\end{aligned} \tag{4.27}$$

For the percentiles, with $p = 90 - 95$, we have $[L_{si,p}]^{(1)} \cong 2\ [E[L_{si}]^{(1)}]$ and, therefore, it is

$$[L_{si,p}]^{(1)} = [\,1.18\ 0.90\ 1.44\ 0.98\,]\ \text{(veh)} \tag{4.28}$$

State 2

To calculate the circulating flows $[Q_{ei}^{(2)}]$ to be introduced in the capacity formula, we must now take into account the number of vehicles $[E[L_{si}]^{(1)}]$ in the system (waiting at legs) determined for the previous condition (vector (4.27)) that enter the circulatory roadway during the interval $T_2 = 30$ min, and these vehicles must, therefore, be added to the vehicles that form the demand flow $[Q_{ei}^{(2)}]$.

In conclusion, $[Q_{ei}^{(2)}]$ (Eq. (4.12)) and $[E[L_{si}]^{(1)}]$ (Eq. (4.27)) yield (Eq. (4.4)) the volumes (expressed in pcu/h)

$$\begin{aligned} Q_{e1}^{(2)*} &= 455 + 0.59 \cdot \tfrac{60}{30} = 456 \\ Q_{e2}^{(2)*} &= 460 + 0.45 \cdot \tfrac{60}{30} = 461 \\ Q_{e3}^{(2)*} &= 433 + 0.72 \cdot \tfrac{60}{30} = 434 \\ Q_{e4}^{(2)*} &= 420 + 0.49 \cdot \tfrac{60}{30} = 421 \end{aligned} \tag{4.29}$$

that is

$$[Q_{ei}^{(2)*}] = \begin{bmatrix} 456\ 461\ 434\ 421 \end{bmatrix} \text{ (pcu/h)} \tag{4.30}$$

$[Q_{ei}^{(2)*}]$ actually coincides with the vector $[Q_{ei}^{(2)}]$ (vector (4.12)).

With the elements of vector (4.30) and matrix (4.12), Eq. (1.12) of Chap. 1 yields the circle flows that follows

$$[Q_{ci}^{(2)*}] = \begin{bmatrix} 466\ 464\ 436\ 334 \end{bmatrix} \text{ (pcu/h)} \tag{4.31}$$

With these values, we can calculate the entry capacity determinations during $T_2 = 30$ min using Eq. (2.12) of Chap. 2.

$$[C_i^{(2)}] = \begin{bmatrix} 842\ 843\ 866\ 950 \end{bmatrix} \text{ (pcu/h)} \tag{4.32}$$

The traffic intensity vector $\rho_i^{(2)} = Q_{ei}^{(2)}/C_i^{(2)}$ is

$$[\rho_i^{(2)}] = [\rho_1^{(2)}\ \rho_2^{(2)}\ \rho_3^{(2)}\ \rho_4^{(2)}] = [0.54\ 0.55\ 0.50\ 0.44] \tag{4.33}$$

With capacity (4.32) and traffic intensities (4.33), from Eq. (4.1), we have the values for each time interval $\vartheta_i^{(2)}$

$$
\begin{aligned}
\vartheta_1^{(2)} &= \frac{1}{C_1^{(2)}\cdot(1-\sqrt{\rho_1^{(2)}})^2} = \frac{1}{(842/3600)\cdot(1-\sqrt{0.54})^2} = 60.8\ \text{s} \\
\vartheta_2^{(2)} &= \frac{1}{C_2^{(2)}\cdot(1-\sqrt{\rho_2^{(2)}})^2} = \frac{1}{(843/3600)\cdot(1-\sqrt{0.55})^2} = 63.9\ \text{s} \\
\vartheta_3^{(2)} &= \frac{1}{C_3^{(2)}\cdot(1-\sqrt{\rho_3^{(2)}})^2} = \frac{1}{(866/3600)\cdot(1-\sqrt{0.50})^2} = 48.5\ \text{s} \\
\vartheta_4^{(2)} &= \frac{1}{C_4^{(2)}\cdot(1-\sqrt{\rho_4^{(2)}})^2} = \frac{1}{(950/3600)\cdot(1-\sqrt{0.44})^2} = 33.4\ \text{s}
\end{aligned}
\quad (4.34)
$$

All the values of $\vartheta_i^{(2)}$ obtained are much smaller than the observation interval $T_2 = 30$ min $= 1800$ s.

The assumption of instantaneous transition from stage 2 to stage 3 that we made at the beginning of this worked example is, therefore, plausible.

In other words, the interval $T_2 = 30$ min is sufficiently wide to allow the roundabout to rapidly reach steady-state conditions and preserve them in stage 2.

The averages of the times spent in the system (in s) $E[w_{si}]^{(2)}$ and the number of vehicles in the system $E[L_{si}]^{(2)}$ can be obtained from the relationships shown in Table 3.3 of Chap. 3 using the values of capacity and traffic intensity described in Eqs. (4.32) and (4.33), respectively

$$
\begin{aligned}
[E[w_{si}]^{(2)}] &= [E[w_{s1}]^{(2)}\ E[w_{s2}]^{(2)}\ E[w_{s3}]^{(2)}\ E[w_{s4}]^{(2)}] = \\
&= [\,9.30\ 9.49\ 8.31\ 6.77\,]\ (\text{s})
\end{aligned}
\quad (4.35)
$$

$$
\begin{aligned}
[E[L_{si}]^{(2)}] &= [E[L_{s1}]^{(2)}\ E[L_{s2}]^{(2)}\ E[L_{s3}]^{(2)}\ E[L_{s4}]^{(2)}] = \\
&= [\,1.17\ 1.22\ 1.00\ 0.79\,]\ (\text{veh})
\end{aligned}
\quad (4.36)
$$

For the percentiles, with $p = 90 - 95$, we have $[L_{si,p}]^{(2)} \cong 2\ [E[L_{si}]^{(2)}]$, and, therefore, it is

$$
[L_{si,p}]^{(2)} = [\,2.34\ 2.44\ 2.00\ 1.58\,]\ (\text{veh}) \quad (4.37)
$$

4.1.2 Second Worked Example

Assume that the roundabout described in the previous example evolves again through three states, with the first and the third associated with traffic demand (pcu/h) given by

$$[Q_{ei}^{(0)}] = [Q_{ei}^{(2)}] = \left[590\ 540\ 500\ 530 \right]$$

$$P_{O/D}^{(0)} \equiv P_{O/D}^{(2)} \equiv \begin{bmatrix} 0 & 0.40 & 0.40 & 0.20 \\ 0.35 & 0 & 0.50 & 0.15 \\ 0.15 & 0.30 & 0 & 0.55 \\ 0.40 & 0.40 & 0.20 & 0 \end{bmatrix}$$

$$M_{O/D}^{(0)} \equiv M_{O/D}^{(2)} \equiv \begin{bmatrix} 0 & 272 & 272 & 136 \\ 210 & 0 & 300 & 90 \\ 110 & 219 & 0 & 402 \\ 220 & 220 & 110 & 0 \end{bmatrix}$$

This demand is applied for a duration that can be considered, for practical aims, infinite.

The second state, the duration of which is $T_1 = 10$ min, is connected to the matrix $P_{O/D}^{(1)}$ (4.10) and the vector $[Q_{ei}^{(1)}]$

$$[Q_{ei}^{(1)}] = [\, 638\ 590\ 660\ 680 \,]\, (\text{pcu/h}) \tag{4.38}$$

that is to matrix $M_{O/D}^{(1)}$

$$M_{O/D}^{(1)} \equiv \begin{bmatrix} 0 & 159 & 230 & 249 \\ 171 & 0 & 218 & 201 \\ 218 & 191 & 0 & 251 \\ 211 & 238 & 231 & 0 \end{bmatrix} (\text{pcu/h}) \tag{4.39}$$

Also in this example, to simplify the procedure, we assume that, in the flows under examination, the percentages of vehicles different from passenger cars are completely negligible, so that the values expressed in pcu/h and pcu coincide with those expressed in veh/h and veh. Again, we use the capacity formula developed by Brilon-Wu (See Sect. 2.1.5).

State 0

By the same calculation criteria used for the homologous condition of the worked example described in Sect. 4.1.1, we have

$$[Q_{ci}^{(0)}] = [\, 468\ 460\ 388\ 414 \,]\, (\text{puc/h})$$

$$[C_i^{(0)}] = [\, 840\ 847\ 905\ 884 \,]\, (\text{puc/h})$$

$$[\rho_i^{(0)}] = [\, 0.70\ 0.64\ 0.55\ 0.60 \,]$$

$$[E[w_{si}]^{(0)}] = [\, 14.28\ 11.81\ 8.84\ 10.18 \,]\, (\text{s})$$

$$[E[L_{si}]^{(0)}] = [\, 2.33\ 1.78\ 1.22\ 1.50 \,]\, (\text{veh})$$

$$[L_{si,p}]^{(0)} = [\, 4.66\ 3.56\ 2.44\ 3.00 \,]\, (\text{veh})$$

State 1

To calculate the circulating flows $[Q_{ei}^{(1)}]$, we take into account the number of vehicles $[E[L_{si}]^{(0)}]$, determined for the previous condition, that enter the circulatory roadway during the interval $T_1 = 10$ min and that must, therefore, be added to the vehicles that form the demand flow $[Q_{ei}^{(1)}]$.

Finally, $[Q_{ei}^{(1)}]$ (Eq. (4.38)) and $[E[L_{si}]^{(0)}]$ yield, on the basis of Eqs. (4.4) and (4.5), the volumes (expressed in pcu/h)

$$\begin{aligned} Q_{e1}^{(1)*} &= \left(638 \cdot \tfrac{10}{60} + 2.33\right) \cdot \tfrac{60}{10} = 652 \\ Q_{e2}^{(1)*} &= \left(590 \cdot \tfrac{10}{60} + 1.78\right) \cdot \tfrac{60}{10} = 601 \\ Q_{e3}^{(1)*} &= \left(660 \cdot \tfrac{10}{60} + 1.22\right) \cdot \tfrac{60}{10} = 667 \\ Q_{e4}^{(1)*} &= \left(680 \cdot \tfrac{10}{60} + 1.50\right) \cdot \tfrac{60}{10} = 689 \end{aligned} \tag{4.40}$$

that is

$$[Q_{ei}^{(1)*}] = \left[652\ 601\ 667\ 689\right] \text{ (pcu/h)} \tag{4.41}$$

With the elements of vector (4.41) and matrix $P_{O/D}^{(1)}$ (matrix 4.10), Eq. (1.12) of Chap. 1 yields the circle flows that follow

$$[Q_{ei}^{(1)*}] = \left[669\ 723\ 633\ 589\right] \text{ (pcu/h)} \tag{4.42}$$

With these values, we can calculate entry capacities during $T_1 = 10$ min from Eq. (2.12) of Chap. 2.

$$[C_i^{(1)}] = \left[683\ 643\ 711\ 745\right] \text{ (pcu/h)} \tag{4.43}$$

With vectors (4.38) and (4.43) we calculate the traffic intensity vector $\rho_i{}^{(1)} = Q_{ei}{}^{(1)}/C_i{}^{(1)}$

$$[\rho_i^{(1)}] = [\rho_1^{(1)}\ \rho_2^{(1)}\ \rho_3^{(1)}\ \rho_4^{(1)}] = [0.93\ 0.92\ 0.93\ 0.91] \tag{4.44}$$

With the capacities and traffic intenities described in Eqs. (4.43) and (4.44), respectively, from Eq. (4.1) we have the values for each time interval $\vartheta_i{}^{(1)}$

$$\begin{aligned}
\vartheta_1^{(1)} &= \frac{1}{C_1^{(1)}\cdot(1-\sqrt{\rho_1^{(1)}})^2} = \frac{1}{(683/3600)\cdot(1-\sqrt{0.93})^2} = 4148\,\mathrm{s} = 69.1\,\mathrm{min}\\
\vartheta_2^{(1)} &= \frac{1}{C_2^{(1)}\cdot(1-\sqrt{\rho_2^{(1)}})^2} = \frac{1}{(643/3600)\cdot(1-\sqrt{0.92})^2} = 3359\,\mathrm{s} = 56.0\,\mathrm{min}\\
\vartheta_3^{(1)} &= \frac{1}{C_3^{(1)}\cdot(1-\sqrt{\rho_3^{(1)}})^2} = \frac{1}{(711/3600)\cdot(1-\sqrt{0.93})^2} = 3988\,\mathrm{s} = 66.5\,\mathrm{min}\\
\vartheta_4^{(1)} &= \frac{1}{C_4^{(1)}\cdot(1-\sqrt{\rho_4^{(1)}})^2} = \frac{1}{(745/3600)\cdot(1-\sqrt{0.91})^2} = 2279\,\mathrm{s} = 38.0\,\mathrm{min}
\end{aligned} \tag{4.45}$$

All the values of $\vartheta_i^{(1)}$ obtained are much greater than the observation interval $T_1 = 10$ min $= 600$ s.

The assumption of instantaneous transition from stage 0 to stage 1 is, therefore, not plausible.

In other words, the interval $T_1 = 10$ min is not sufficiently wide to allow the roundabout to reach steady-state conditions and preserve them in stage 1.

Therefore, the averages of the times spent and the number of vehicles in the system cannot be obtained from the relationships that are valid for the steady-state conditions. Instead, they must be calculated starting with Eq. (3.142) and with the specifications of Eqs. (3.143)–(3.147), all of which are contained in Chap. 3 (flows and capacities in pcu/h; time T in h).

$$<w_{si}> = \left[\frac{1}{2}\cdot\left(\sqrt{F_i^2 + G_i} - F_i\right) + E_i + \frac{1}{C_i^{(1)}}\right]\cdot 3600\,(\mathrm{s}) \tag{4.46}$$

where the subscript "i" refers to the factors of the second member relative to the generic entry "i:"

$$F_i = \frac{1}{C_i^{(0)} - Q_{ei}^{(0)}}\cdot\left[\frac{T_1}{2}\cdot(C_i^{(1)} - Q_{ei}^{(1)})\cdot z_i + \left(z_i - \frac{h_i}{C_i^{(1)}}\right)\right] + E_i \tag{4.47}$$

$$G_i = \frac{2\cdot T_1\cdot z_i}{C_i^{(0)} - Q_{ei}^{(0)}}\cdot\left[\frac{Q_{ei}^{(1)}}{C_i^{(1)}} - (C_i^{(1)} - Q_{ei}^{(1)})\cdot E_i\right] \tag{4.48}$$

$$E_i = \frac{Q_{ei}^{(0)}}{C_i^{(0)}\cdot(C_i^{(0)} - Q_{ei}^{(0)})} \tag{4.49}$$

$$h_i = C_i^{(1)} - C_i^{(0)} + Q_{ei}^{(0)} \tag{4.50}$$

$$z_i = 1 - \frac{h_i}{Q_{ei}^{(1)}} \tag{4.51}$$

Substituting the values of traffic demand and capacity calculated before for state 0 and state 1 into Eqs. (4.47)–(4.51) for entry 1, we have

$$h_1 = C_1^{(1)} - C_1^{(0)} + Q_{e1}^{(0)} = 683 - 840 + 590 = 433$$

$$z_1 = 1 - \frac{h_1}{Q_{e1}^{(1)}} = 1 - \frac{433}{638} = 0.3213166$$

$$E_1 = \frac{Q_{e1}^{(0)}}{C_1^{(0)} \cdot (C_1^{(0)} - Q_{e1}^{(0)})} = \frac{590}{840 \cdot (840 - 590)} = 0.0028095$$

$$F_1 = \frac{1}{C_1^{(0)} - Q_{e1}^{(0)}} \cdot \left[\frac{T_1}{2} \cdot (C_1^{(1)} - Q_{e1}^{(1)}) \cdot z_1 + \left(z_1 - \frac{h_1}{C_1^{(1)}}\right)\right] + E_1 =$$

$$= \frac{1}{840 - 590} \cdot \left[\frac{10}{2} \cdot (683 - 638) \cdot 0.3213166 + \left(0.3213166 - \frac{433}{683}\right)\right]$$

$$+ 0.0028095 = 0.0063787$$

$$G_1 = \frac{2 \cdot T_1 \cdot z_1}{C_1^{(0)} - Q_{e1}^{(0)}} \cdot \left[\frac{Q_{e1}^{(1)}}{C_1^{(1)}} - (C_1^{(1)} - Q_{e1}^{(1)}) \cdot E_1\right] =$$

$$= \frac{2 \cdot 10 \cdot 0.3213166}{840 - 590} \cdot \left[\frac{638}{683} - (683 - 638) \cdot 0.0028095\right] = 0.0003460$$

and, therefore,

$$<w_{s1}>^{(1)} = \left[\frac{1}{2} \cdot \left(\sqrt{F_1^2 + G_1} - F_1\right) + E_1 + \frac{1}{C_1^{(1)}}\right] \cdot 3600 =$$

$$= \left[\frac{1}{2} \cdot \left(\sqrt{(0.0063787)^2 + 0.0003460} - 0.0063787\right) + 0.0028095 + \frac{1}{683}\right] \cdot$$

$$3600 = 39.3\,\text{s}$$

Repeating the calculation just performed for entry 1, for the remaining three entries we have

$$<w_{s2}>^{(1)} = 36.5\ \text{s}$$

$$<w_{s3}>^{(1)} = 34.3\ \text{s}$$

$$<w_{s4}>^{(1)} = 31.9\ \text{s}$$

To determine the 95th percentile of L_{si} we use Eq. (3.150) from Chap. 3, and, thus, we have

$$L_{s1.95} = \frac{C_1^{(1)} \cdot T}{4} \cdot \left[\rho_1^{(1)} - 1 + \sqrt{(1 - \rho_1^{(1)})^2 + \frac{8 \cdot \rho_1^{(1)}}{C_1^{(1)} \cdot T} \cdot 3} \right] =$$

$$= \frac{683 \cdot 0.167}{4} \cdot \left[0.93 - 1 + \sqrt{(1 - 0.93)^2 + \frac{8 \cdot 0.93}{683 \cdot 0.167} \cdot 3} \right] = 10.8 \text{ veh}$$

$$L_{s2.95} = 10.2 \text{ veh}$$

$$L_{s3,95} = 10.9 \text{ veh}$$

$$L_{s2,95} = 10.5 \text{ veh}$$

Applying in any case, to stage 1, the relationship at steady-state conditions (See Table 3.3 of Chap. 3,) we have the following values

$$[E[w_{si}]^{(1)}] = [\,75.3\ 70.0\ 72.4\ 53.7\,]$$
$$[L_{si,95}]^{(1)} = [\,26.6\ 23.0\ 26.6\ 20.2\,]\ (\text{veh})$$

by which we overestimate the determinations of the state variables under examination.

State 2

To calculate the circulating flows $[Q_{ci}^{(2)}]$ to be introduced into the capacity formula, we must take into account the number of vehicles $< L_{si} >^{(1)}$ in the system (waiting at legs) at the end of the previous period T_1. However, since the value of T_2 is in this example assumed to be infinite, the entering flows $[Q_{ei}^{(2)*}]$ coincide with the demand flows (Eq. (4.4)).

In addition, having assumed that traffic demand is equal to $[Q_{ei}^{(0)}] = [Q_{ei}^{(2)}]$ and $M_{O/D}^{(0)} = M_{O/D}^{(2)}$, it follows that, under this condition, the determinations of $[E[w_{si}]^{(2)}]$ and $[E[L_{si}]^{(2)}]$ coincide with those of state 0.

In the end, we have

$$[E[w_{si}]^{(2)}] = [\,14.28\ 11.81\ 8.84\ 10.18\,]\,(\text{s})$$
$$[E[L_{si}]^{(2)}] = [\,2.33\ 1.78\ 1.22\ 1.50\,]\,(\text{veh})$$
$$[L_{si,95}]^{(2)} = [\,4.66\ 3.56\ 2.44\ 3.00\,]\,(\text{veh})$$

4.2 State Evolution with Saturated or Oversaturated Entries

In the case of evolution of the system between conditions with saturated or oversaturated entries, we use time-dependent formulas since there are no steady-state conditions. For each entry, the behavior of traffic demand with time is known (assigned, for example, as in the form in Sect. 1.1).

The beginning and the end of observation interval T are selected in correspondence with the system under steady-state conditions. T is divided into subintervals T_k, during which $Q_{ei}(t)$ is constant (See Fig. 4.1).

The calculation of the state variables of the system is performed with the following procedure:

- we select a capacity formula for the roundabout under examination;
- for the initial steady-state condition, we determine, with the criteria described in Sect. 1.1, the circle flows, the exiting flows, and the disturbing volumes to be introduced into the capacity formula selected in correspondence with each leg;
- we calculate the first capacity value for each entry "i";
- then, for each entry, we use the relationships described in Sect. 3.1 to determine the average time spent $[E[w_{si}]^{(0)}]$ and the number of vehicles in the system $[E[L_{si}]^{(0)}]$ (and/or the waiting time in the queue and the queue length if we also wish to use these state variables) at statistical equilibrium;
- at each successive calculation step k (with the respective interval T_k), we determine for each entry:
  - the demand value $Q_{ei}^{(k)*}$, on the basis of Eq. (4.4), as a function of the value $Q_{ei}^{(k)}$ of traffic demand during T_k and of the number of vehicles in the system at the end of the previous period T_{k-1};
  - with the $Q_{ei}^{(k)*}$ relative to each entry, the circle flows, the exiting flows, and the conflicting flows to be introduced into the selected capacity formula;
  - the capacities $C_i^{(k)}$ at each entry (in case of saturation or oversaturation of one or more of the entries using the procedure illustrated in Sect. 2.5);
  - the traffic intensities (degrees of saturation) $\rho_i^{(k)} = Q_{ei}^{(k)}/C_i^{(k)}$;
  - the number of users in the system $<L_{si}>^{(k)}$ at the end of the interval T_k ((Eq. (3.103) of Chap. 3) using, for each entry "i" during the interval T_k, Eqs. (3.104) and (3.105) of Chap. 3 with:

$$\rho = \rho_i^{(k)}; \quad C = C_i^{(k)}; \quad T = T_k; \quad L_{s0} = <L_{si}>^{(k-1)}$$

thus, obtaining at step k

$$<L_{si}>^{(k)} = \frac{1}{2}\left(\sqrt{(A_i^{(k)})^2 + B_i^{(k)}} - A_i^{(k)}\right) \tag{4.52}$$

with

$$A_i^{(k)} = (1 - \rho_i^{(k)}) \cdot C_i^{(k)} \cdot T_k + 1 - <L_{si}>^{(k-1)}) \tag{4.53}$$

$$B_i = 4(<L_{si}>^{(k-1)} + \rho_i^{(k)} \cdot C_i^{(k)} \cdot T_k) \tag{4.54}$$

We repeat that the value L_{s0} of Eq. (3.104) of Chap. 3 evidently coincides with the number of users in the system $<L_{si}>^{(k-1)}$ at the end of interval T_{k-1} previous to the one considered in the calculation (at the beginning of the calculation procedure we have $k = 1$ and $L_{s0} = E[L_{si}]^{(0)}$);

- the average time spent in the system $<w_{si}>^{(k)}$ during T_k, with Eq. (3.115) of Chap. 3. Equation (3.115) of Chap. 3 is specified for each entry "i" putting into Eqs. (3.116) and (3.117) of Chap. 3:

$$\rho = \rho_i^{(k)}; \quad C = C_i^{(k)}; \quad T = T_k; \quad L_{s0} = <L_{si}>^{(k-1)}$$

thus, we obtain at step k:

$$<w_{si}>^{(k)} = \frac{1}{2}\left(\sqrt{(J_i^{(k)})^2 + M_i^{(k)}} - J_i^{(k)}\right) \tag{4.55}$$

with

$$J_i^{(k)} = \frac{T_k}{2} \cdot (1 - \rho_i^{(k)}) - \frac{1}{C_i^{(k)}} \cdot (<L_{si}>^{(k-1)} + 1) \tag{4.56}$$

$$M_i^{(k)} = \frac{4}{C_i^{(k)}} \cdot \left[\frac{T_k}{2} \cdot (1 - \rho_i^{(k)}) + \frac{1}{2} \cdot \rho_i^{(k)} \cdot T_k\right] \tag{4.57}$$

The procedure ends when the entire system observation interval has been covered. Evidently, if we are interested in the determination of the queue lengths and waiting times in the queue, we must also specify, with the same positions used for Eqs. (3.103)–(3.105) and for Eqs. (3.115)–(3.117), Eqs. (3.107)–(3.109), and Eqs. (3.119)–(3.121) of Chap. 3. So, we must use Eqs. (3.107) and (3.119) from Chap. 3.

To better explain the points illustrated so far, we will now give two calculation examples.

4.2.1 First Worked Example

At the beginning of the computational process, the roundabout that is assumed to have four legs (i = 1, 2, 3, 4) is at steady-state conditions.

Traffic demand has the following percentage composition: heavy vehicles (coaches included) account for 6.5%, and bicycles and motorcycles account for 5% of the total volume Q_{ei}.

This composition remains the same on all legs of the system during the entire observation interval T.

For the transformation from pcu/h and pcu into veh/h and veh, respectively, the coefficient f given by Eq. (4.7) is

$$f = \frac{1}{(1 - 0.065 - 0.05) + 2 \cdot 0.065 + 0.5 \cdot 0.05} = 0.962$$

where we have used the values $\alpha_p = 2$ and $\alpha_{cm} = 0.5$, respectively, for the equivalence in passenger cars.

In addition, the matrix $P_{O/D}$ of the traffic percentages is also assumed to be invariant during the entire T. The variability of demand in time is thus attributable only to the behavior of flows $Q_{ei}^{(k)}$ during the sub-intervals T_k (See Fig. 4.3).

To give an example, we follow the evolution process relative to entry 2 (The procedure can be repeated for the other entries).

From the calculation of disturbing flows $Q_{di}^{(0)}$ on the basis of traffic demand ($Q_{ei}^{(0)}$; $P_{O/D}$), with the selected capacity formula we have, for entry 2, the capacity value $C_2^{(0)} = 1055$ pcu/h $= 1055 \cdot 0.962 = 1015$ veh/h.

As $Q_{ei}^{(0)} = 895$ pcu/h (See Fig. 4.3), for the traffic intensity $\rho_2^{(0)} = 895/\ 1055 = 0.848$ and, therefore, for the number of vehicles $E[L_{s2}]^{(0)}$ and the time spent $E[w_{s2}]^{(0)}$ in the system (See Table 3.3 of Chap. 3), assuming Poissonian arrivals and exponential service times, we have:

$$E[w_{s2}]^{(0)} = 1/[(1015/3600) \cdot (1 - 0.848)] = 23.3 \text{ s}$$
$$E[L_{s2}]^{(0)} = 0.848/(1 - 0.848) = 5.6 \text{ veh}$$

Similarly, since we know $Q_{ei}^{(0)}$ and have determined $C_i^{(0)}$ for the other three entries, we have the initial values $[E[w_{si}]^{(0)}]$ and $E[L_{si}]^{(0)}]$.

Demand evolution is given for intervals of $T_k = 10$ min.

In step 1, we calculate (Eq. (4.4)), for each entry, the increase in demand caused by the number of vehicles waiting in the system relative to the previous state; to do so, the $[E[L_{si}]^{(0)}]$ must be converted from veh to pcu, once again by using the coefficient $f = 0.962$.

Thus, for entry 2, since $Q_{e2}^{(1)} = 1034$ pcu/h $= 1034 \cdot 0.962 = 1000$ veh/h and $E[L_{s2}]^{(0)} = 5.6/0.962 = 5.82$ pcu/h, we have

$$Q_{e2}^{(1)*} = 1034 + 5.82 \cdot \frac{60}{10} = 1069 \text{pcu/h}$$

Since we know the demand $[Q_{ei}^{(1)*}]$ we determine, for each entry, the disturbing flows and the capacity values $[C_i^{(1)}]$ relative to calculation step 1, i.e., for interval T_1.

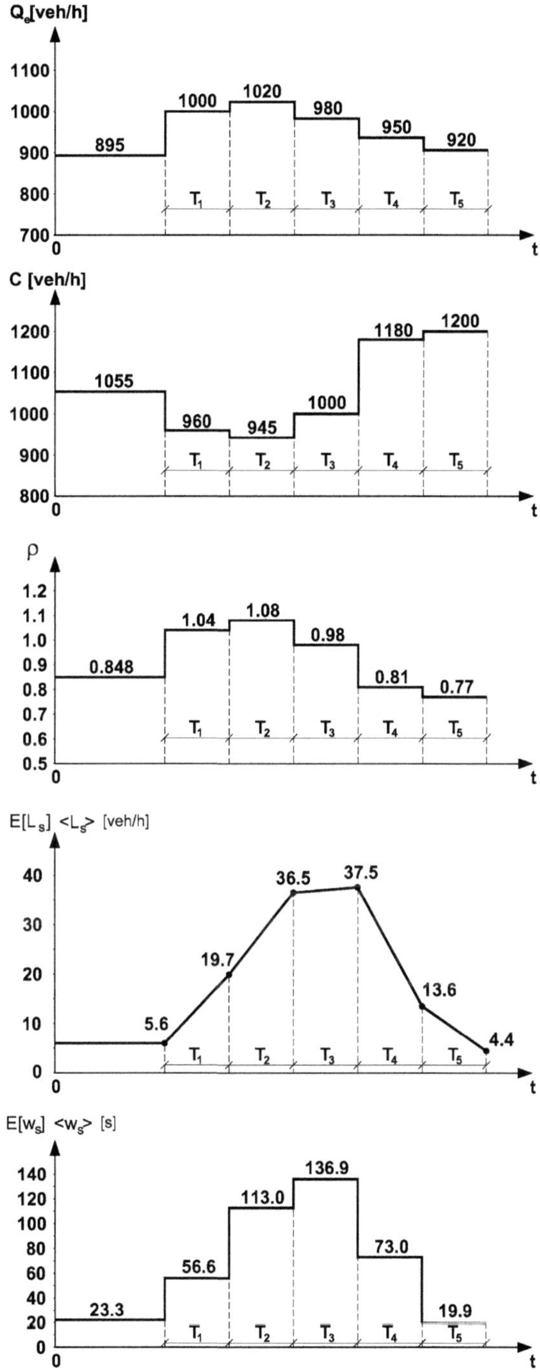

Fig. 4.3 Time evolution of traffic demand, capacity, traffic intensity, and state variables for entry 2

For $C_2^{(1)}$, our calculation yields $C_2^{(1)} = 998$ pcu/h. The entry is oversaturated, and, therefore, we used the computational procedure described in Sect. 2.5 for the determination of capacity.

To determine $<w_{s2}>^{(1)}$ and $<L_{s2}>^{(1)}$, we convert $C_2^{(1)}$ from pcu/h to veh/h. Thus, we have

$$C_2^{(1)} = 998 \cdot 0.962 = 960 \text{ veh/h}$$

and we evaluate traffic intensity during the period T_1 that is equal to

$$\rho_2^{(1)} = Q_{e2}^{(1)}/C_2^{(1)} = 1000/960 = 1.04$$

Putting these values into Eqs. (4.53) and (4.54), we have

$$\begin{aligned} A_2^{(1)} &= (1 - \rho_2^{(1)}) \cdot C_2^{(1)} \cdot T_1 + 1 - <L_{s2}>^{(0)} \\ &= (1 - 1.04) \cdot 960/3600 \cdot 600 + 1 - 5.6 = -11.00 \end{aligned}$$

$$\begin{aligned} B_2^{(1)} &= 4 \cdot (<L_{s2}>^{(0)} + \rho_2^{(1)} \cdot C_2^{(1)} \cdot T_1) \\ &= 4 \cdot (5.6 + 1.04 \cdot 960/3600 \cdot 600) = 688.00 \end{aligned}$$

and, therefore, with Eq. (4.52), we have

$$\begin{aligned} <L_{s2}>^{(1)} &= \frac{1}{2} \cdot \left(\sqrt{(A_2^{(1)})^2 + B_2^{(1)}} - A_2^{(1)} \right) = \\ &= \frac{1}{2} \cdot \left(\sqrt{(-11.00)^2 + 688.00} + 11.00 \right) = 19.7 \text{ veh} \end{aligned}$$

By specifying Eqs. (4.56) and (4.57), we have

$$\begin{aligned} J_2^{(1)} &= \frac{T_1}{2} \cdot (1 - \rho_2^{(1)}) - \frac{1}{C_2^{(1)}} \cdot (<L_{s2}>^{(0)} + 1) = \\ &= \frac{600}{2} \cdot (1 - 1.04) - \frac{1}{960/3600} \cdot (5.6 + 1) = -36.75 \end{aligned}$$

$$\begin{aligned} M_2^{(1)} &= \frac{4}{C_2^{(1)}} \cdot \left[\frac{T_1}{2} \cdot (1 - \rho_2^{(1)}) + \frac{1}{2} \cdot \rho_2^{(1)} \cdot T_1 \right] = \\ &= \frac{4}{960/3600} \cdot \left[\frac{600}{2} \cdot (1 - 1.04) + \frac{1}{2} \cdot 1.04 \cdot 600 \right] = 4500.00 \end{aligned}$$

and, therefore, from Eq. (4.55), we have

$$\begin{aligned} <w_{s2}^{(1)}> &= \frac{1}{2} \cdot \left(\sqrt{(J_2^{(1)})^2 + M_2^{(1)}} - J_2^{(1)} \right) = \\ &= \frac{1}{2} \cdot \left(\sqrt{(-36.75)^2 + 4500.00} + 36.75 \right) = 56.6 \text{ s} \end{aligned}$$

In step 2, we calculate (Eq. (4.4)), for each entry, the increase in demand caused by the number of vehicles in the system relative to the previous state; to do so, the $<L_{si}>^{(1)}$ must be converted from veh to pcu, once again using the coefficient f = 0.962.

Therefore, for entry 2, since $Q_{e2}^{(2)} = 1060$ pcu/h $= 1000 \cdot 0.962 = 1020$ veh/h and $<L_{s2}>^{(1)} = 19.7/0.962 = 20.48$ pcu/h, we have

$$Q_{e2}^{(2)*} = 1060 + 20.48 \cdot \frac{60}{10} = 1183 \text{ pcu/h}$$

Since we know the demand $[Q_{ei}^{(2)*}]$, we determine, for each entry, the disturbing flows and then the capacity values $[C_i^{(2)}]$ relative to calculation step 2, that is, for the interval T_2.

For $C_2^{(2)}$, our calculation yields $C_2^{(2)} = 982$ pcu/h (the entry is oversaturated, and, thus, for the calculation of capacity, we used the computational procedure described in Sect. 2.5).

To determine $<w_{s2}>^{(2)}$ and $< L_{s2} >^{(2)}$, we convert $C_2^{(2)}$ from pcu/h to veh/h. Thus, we have

$$C_2^{(2)} = 982 \cdot 0.962 = 945 \text{ veh/h}$$

and we evaluate traffic intensity during period T_2, which is equal to

$$\rho_2^{(2)} = Q_{e2}^{(2)}/C_2^{(2)} = 1020/945 = 1.08$$

Putting these values into Eqs. (4.53) and (4.54), we have

$$\begin{aligned} A_2^{(2)} &= (1 - \rho_2^{(2)}) \cdot C_2^{(2)} \cdot T_2 + 1 - < L_{s2} >^{(1)} = \\ &= (1 - 1.08) \cdot 945/3600 \cdot 600 + 1 - 19.7 = -31.30 \end{aligned}$$

$$\begin{aligned} B_2^{(2)} &= 4 \cdot (<L_{s2}>^{(1)} + \rho_2^{(2)} \cdot C_2^{(2)} \cdot T_2) = \\ &= 4 \cdot (19.7 + 1.08 \cdot 945/3600 \cdot 600) = 759.20 \end{aligned}$$

and, therefore, with Eq. (4.52), we have

$$<L_{s2}>^{(2)} = \tfrac{1}{2} \cdot \left(\sqrt{(A_2^{(2)})^2 + B_2^{(2)}} - A_2^{(2)}\right) =$$

$$= \tfrac{1}{2} \cdot \left(\sqrt{(-31.30)^2 + 759.20} + 31.30\right) = 36.5 \text{ veh}$$

By specifying Eqs. (4.56) and (4.57), we have

$$\begin{aligned}
J_2^{(2)} &= \tfrac{T_2}{2}\cdot(1-\rho_2^{(2)}) - \tfrac{1}{C_2^{(2)}}\cdot(<L_{s2}>^{(1)}+1) = \\
&= \tfrac{600}{2}\cdot(1-1.08) - \tfrac{1}{945/3600}\cdot(19.7+1) = -102.86 \\
M_2^{(2)} &= \tfrac{4}{C_2^{(2)}}\cdot\left[\tfrac{T_2}{2}\cdot(1-\rho_2^{(2)}) + \tfrac{1}{2}\cdot\rho_2^{(2)}\cdot T_2\right] = \\
&= \tfrac{4}{945/3600}\cdot\left[\tfrac{600}{2}\cdot(1-1.08) + \tfrac{1}{2}\cdot 1.08\cdot 600\right] = 4571.43
\end{aligned}$$

and therefore, from Eq. (4.55), we have

$$\begin{aligned}
<w_{s2}^{(2)}> &= \tfrac{1}{2}\cdot\left(\sqrt{(J_2^{(2)})^2 + M_2^{(2)}} - J_2^{(2)}\right) = \\
&= \tfrac{1}{2}\cdot\left(\sqrt{(-102.86)^2 + 4571.43} + 102.86\right) = 113.0\ \text{s}
\end{aligned}$$

In step 3, we calculate (Eq. (4.4)), for each entry, the increase in demand caused by the number of vehicles in the system relative to the previous state; to do so, the $<L_{si}>^{(2)}$ must be converted from veh to pcu, once again using the coefficient $f = 0.962$.

Therefore, for entry 2, since $Q_{e2}^{(3)} = 1019$ pcu/h $= 1019 \cdot 0.962 = 980$ veh/h and $<L_{s2}>^{(2)} = 36.5/0.962 = 37.94$ pcu/h, we have

$$Q_{e2}^{(3)*} = 1019 + 37.94\cdot\frac{60}{10} = 1247\ \text{pcu/h}$$

Since we know the demand $[Q_{ei}^{(3)*}]$, we determine, for each entry, the disturbing flows and then the capacity values $[C_i^{(3)}]$ relative to calculation step 3, that is, for the interval T_3.

For $C_2^{(3)}$ our calculation yields $C_2^{(3)} = 1040$ pcu/h (other two entries are oversaturated, and, thus, for the calculation of capacity, we used the computational procedure described in Sect. 2.5).

To determine $<w_{s2}>^{(3)}$ and $<L_{s2}>^{(3)}$, we convert $C_2^{(3)}$ from pcu/h to veh/h. Thus, we have

$$C_2^{(3)} = 1040\cdot 0.962 = 1000\ \text{veh/h}$$

and we evaluate traffic intensity during the period T_3 as:

$$\rho_2^{(3)} = Q_{e2}^{(3)}/C_2^{(3)} = 980/1000 = 0.98$$

Putting these values into Eqs. (4.53) and (4.54), we have

$$\begin{aligned}
A_2^{(3)} &= (1-\rho_2^{(3)})\cdot C_2^{(3)}\cdot T_3 + 1 - <L_{s2}>^{(2)} = \\
&= (1-0.98)\cdot 1000/3600\cdot 600 + 1 - 36.5 = -32.17 \\
B_2^{(3)} &= 4\cdot(<L_{s2}>^{(2)} + \rho_2^{(3)}\cdot C_2^{(3)}\cdot T_3) = \\
&= 4\cdot(36.5 + 0.98\cdot 1000/3600\cdot 600) = 799.33
\end{aligned}$$

and, therefore, with Eq. (4.52), we have

$$<L_{s2}>^{(3)} = \tfrac{1}{2} \cdot \left(\sqrt{(A_2^{(3)})^2 + B_2^{(3)}} - A_2^{(3)}\right) =$$
$$= \tfrac{1}{2} \cdot \left(\sqrt{(-32.17)^2 + 799.33} + 32.17\right) = 37.5 \text{ veh}$$

By specifying Eqs. (4.56) and (4.57), we have

$$J_2^{(3)} = \tfrac{T_3}{2} \cdot (1 - \rho_2^{(3)}) - \tfrac{1}{C_2^{(3)}} \cdot (<L_{s2}>^{(2)} + 1) =$$
$$= \tfrac{600}{2} \cdot (1 - 0.98) - \tfrac{1}{1000/3600} \cdot (36.5 + 1) = -129.00$$
$$M_2^{(3)} = \tfrac{4}{C_2^{(3)}} \cdot \left[\tfrac{T_3}{2} \cdot (1 - \rho_2^{(3)}) + \tfrac{1}{2} \cdot \rho_2^{(3)} \cdot T_3\right] =$$
$$= \tfrac{4}{1000/3600} \cdot \left[\tfrac{600}{2} \cdot (1 - 0.98) + \tfrac{1}{2} \cdot 0.98 \cdot 600\right] = 4320.00$$

and, therefore, from Eq. (4.55), we have

$$<w_{s2}^{(3)}> = \tfrac{1}{2} \cdot \left(\sqrt{(J_2^{(3)})^2 + M_2^{(3)}} - J_2^{(3)}\right) =$$
$$= \tfrac{1}{2} \cdot \left(\sqrt{(-129.00)^2 + 4320.00} + 129.00\right) = 136.9 \text{ s}$$

In step 4, we calculate (Eq. (4.4)), for each entry, the increase in demand caused by the number of vehicles in the system relative to the previous state; to do so, the $<L_{si}>^{(3)}$ must be converted from veh to pcu, once again using the coefficient f = 0.962.

Thus, for entry 2, since $Q_{e2}^{(4)} = 988$ pcu/h = 988 · 0.962 = 950 veh/h and $<L_{s2}>^{(3)} = 37.5/0.962 = 38.98$ pcu/h, we have

$$Q_{e2}^{(4)*} = 988 + 38.98 \cdot \frac{60}{10} = 1222 \text{ pcu/h}$$

Since we know the demand $[Q_{ei}^{(4)*}]$, we determine, for each entry, the disturbing flows and then the capacity values $[C_i^{(4)}]$ relative to calculation step 4, that is, for the interval T_4.

For $C_2^{(4)}$, our calculation yields $C_2^{(4)} = 1{,}227$ pcu/h (other two entries are oversaturated, and, thus, for the calculation of capacity, we used the computational procedure described in Sect. 2.5.)

To determine $<w_{s2}>^{(4)}$ and $<L_{s2}>^{(4)}$, we convert $C_2^{(4)}$ from pcu/h to veh/h. Thus, we have

$$C_2^{(4)} = 1227 \cdot 0.962 = 1180 \text{ veh/h}$$

and we evaluate traffic intensity during the period T_4 as:

$$\rho_2^{(4)} = Q_{e2}^{(4)}/C_2^{(4)} = 950/1180 = 0.81$$

Putting these values into Eqs. (4.53) and (4.54), we have

$$\begin{aligned} A_2^{(4)} &= (1 - \rho_2^{(4)}) \cdot C_2^{(4)} \cdot T_4 + 1 - <L_{s2}>^{(3)} = \\ &= (1 - 0.81) \cdot 1180/3600 \cdot 600 + 1 - 37.5 = 0.87 \\ B_2^{(4)} &= 4 \cdot (<L_{s2}>^{(3)} + \rho_2^{(4)} \cdot C_2^{(4)} \cdot T_4) = \\ &\quad 4 \cdot (37.5 + 0.81 \cdot 1180/3600 \cdot 600) = 787.20 \end{aligned}$$

and, therefore, with Eq. (4.52), we have

$$\begin{aligned} <L_{s2}>^{(4)} &= \tfrac{1}{2} \cdot \left(\sqrt{(A_2^{(4)})^2 + B_2^{(4)}} - A_2^{(4)} \right) = \\ &= \tfrac{1}{2} \cdot \left(\sqrt{(0.87)^2 + 787.20} - 0.87 \right) = 13.6 \text{ veh} \end{aligned}$$

By specifying Eqs. (4.56) and (4.57), we have

$$\begin{aligned} J_2^{(4)} &= \tfrac{T_4}{2} \cdot (1 - \rho_2^{(4)}) - \tfrac{1}{C_2^{(4)}} \cdot (<L_{s2}>^{(3)} + 1) = \\ &= \tfrac{600}{2} \cdot (1 - 0.81) - \tfrac{1}{1180/3600} \cdot (37.5 + 1) = -60.46 \\ M_2^{(4)} &= \tfrac{4}{C_2^{(4)}} \cdot \left[\tfrac{T_4}{2} \cdot (1 - \rho_2^{(4)}) + \tfrac{1}{2} \cdot \rho_2^{(4)} \cdot T_4 \right] = \\ &= \tfrac{4}{1180/3600} \cdot \left[\tfrac{600}{2} \cdot (1 - 0.81) + \tfrac{1}{2} \cdot 0.81 \cdot 600 \right] = 3361.02 \end{aligned}$$

and, therefore, from Eq. (4.55), we have

$$\begin{aligned} <w_{s2}^{(4)}> &= \tfrac{1}{2} \cdot \left(\sqrt{(J_2^{(4)})^2 + M_2^{(4)}} - J_2^{(4)} \right) = \\ &= \tfrac{1}{2} \cdot \left(\sqrt{(60.46)^2 + 3361.02} + 60.46 \right) = 73.0 \text{ s} \end{aligned}$$

In step 5, we calculate (Eq. (4.4)), for each entry, the increase in demand caused by the number of vehicles in the system relative to the previous state; to do so, the $<L_{si}>^{(4)}$ must be converted from veh to pcu, once again using the coefficient f = 0.962.

Thus, for entry 2, since $Q_{e2}^{(5)}$ = 956 pcu/h = 988 · 0.962 = 920 veh/h and $<L_{s2}>^{(4)}$ = 13.6/0.962 = 14.14 pcu/h, we have

$$Q_{e2}^{(5)*} = 956 + 14.14 \cdot \frac{60}{10} = 1041 \text{ pcu/h}$$

Since we know the demand $[Q_{ei}^{(5)*}]$, we determine, for each entry, the conflicting flows and then the capacity values $[C_i^{(5)}]$ relative to calculation step 5, that is, for the interval T_5.

For $C_2^{(5)}$ our calculation yields $C_2^{(5)}$ = 1247 pcu/h (other three entries are over-saturated, and, thus, for the calculation of capacity, we used the computational procedure described in Sect. 2.5).

To determine $< w_{s2} >^{(5)}$ and $< L_{s2} >^{(5)}$, we must convert $C_2^{(5)}$ from pcu/h to veh/h. Thus we have

$$C_2^{(5)} = 1247 \cdot 0.962 = 1200 \text{ veh/h}$$

and we evaluate traffic intensity during the period T_4 as:

$$\rho_2^{(5)} = Q_{e2}^{(5)}/C_2^{(5)} = 920/1200 = 0.77$$

Putting these values into Eqs. (4.53) and (4.54), we have

$$\begin{aligned}
A_2^{(5)} &= (1 - \rho_2^{(5)}) \cdot C_2^{(5)} \cdot T_5 + 1 - <L_{s2}>^{(4)} = \\
&= (1 - 0.77) \cdot 1200/3600 \cdot 600 + 1 - 13.6 = 33.40 \\
B_2^{(5)} &= 4 \cdot (<L_{s2}>^{(4)} + \rho_2^{(5)} \cdot C_2^{(5)} \cdot T_5) = \\
&\quad 4 \cdot (13.6 + 0.77 \cdot 1200/3600 \cdot 600) = 670.40
\end{aligned}$$

and, therefore, with Eq. (4.52), we have

$$\begin{aligned}
< L_{s2}>^{(5)} &= \tfrac{1}{2} \cdot \left(\sqrt{(A_2^{(5)})^2 + B_2^{(5)}} - A_2^{(5)} \right) = \\
&= \tfrac{1}{2} \cdot \left(\sqrt{(33.40)^2 + 670.40} - 33.40 \right) = 4.4 \text{ veh}
\end{aligned}$$

By specifying Eqs. (4.56) and (4.57), we have

$$\begin{aligned}
J_2^{(5)} &= \tfrac{T_5}{2} \cdot (1 - \rho_2^{(5)}) - \tfrac{1}{C_2^{(5)}} \cdot (<L_{s2}>^{(4)} + 1) = \\
&= \tfrac{600}{2} \cdot (1 - 0.77) - \tfrac{1}{1200/3600} \cdot (13.6 + 1) = 25.20 \\
M_2^{(5)} &= \tfrac{4}{C_2^{(5)}} \cdot \left[\tfrac{T_5}{2} \cdot (1 - \rho_2^{(5)}) + \tfrac{1}{2} \cdot \rho_2^{(5)} \cdot T_5 \right] = \\
&= \tfrac{4}{1200/3600} \cdot \left[\tfrac{600}{2} \cdot (1 - 0.77) + \tfrac{1}{2} \cdot 0.77 \cdot 600 \right] = 3600.00
\end{aligned}$$

and, therefore, from Eq. (4.55), we have

$$\begin{aligned}
<w_{s2}^{(5)}> &= \tfrac{1}{2} \cdot \left(\sqrt{(J_2^{(5)})^2 + M_2^{(5)}} - J_2^{(5)} \right) = \\
&= \tfrac{1}{2} \cdot \left(\sqrt{(25.20)^2 + 3600.00} - 25.20 \right) = 19.9 \text{ s}
\end{aligned}$$

The results of this worked example are shown in Fig. 4.3, which shows that, by the procedure adopted, it is possible to follow the behavior in time of the development of the queue (increase and clearance) and of the time spent in the system.

4.2.2 Second Worked Example

If traffic demand is given as a time continuous function, the procedure described in Sect. 4.2.1, implemented with the calculation steps T_k of reduced value (e.g., 1 min), allows us to regularly follow the evolution of the system and to obtain the description of the behavior of the intersection under general service conditions.

As an example of what we have just described, using the calculation code 'Roundabout' developed by M. Corradini in Visual Basic®, where the above-mentioned procedure is implemented, the case of a four-legged roundabout was analyzed for traffic demands that have different behaviors for all of the entries. For the first and second legs, the demand is parabolic; for the third and fourth legs, the demand is sinusoidal; and the instants of beginning and end of the peaks and their duration are different for each entry (See Fig. 4.4).

In addition, the time variation of matrix $P_{O/D}$ of the assumed traffic percentages is such that entries 2 and 4 are, during the evolution of the system, always undersaturated (See Fig. 4.5).

Figure 4.4 also shows the relationship $C_i = C_i(t)$ obtained with the capacity calculation performed for the scheme under examination.

With the examination of the behavior of the average number of users and average times spent in the system, described in Figs. 4.6 and 4.7, respectively, it is possible to follow all aspects of the evolution of the roundabout. Therefore, we can evaluate, in particular, the effects that the time variation of traffic demand and, consequently of capacity, have on the waiting phenomena that occur at each entry.

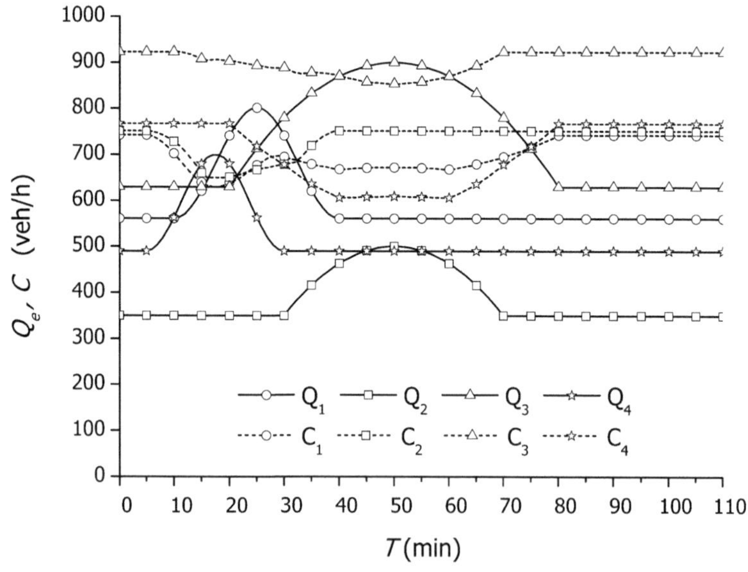

Fig. 4.4 Traffic demand and capacity versus time at the four legs of a roundabout

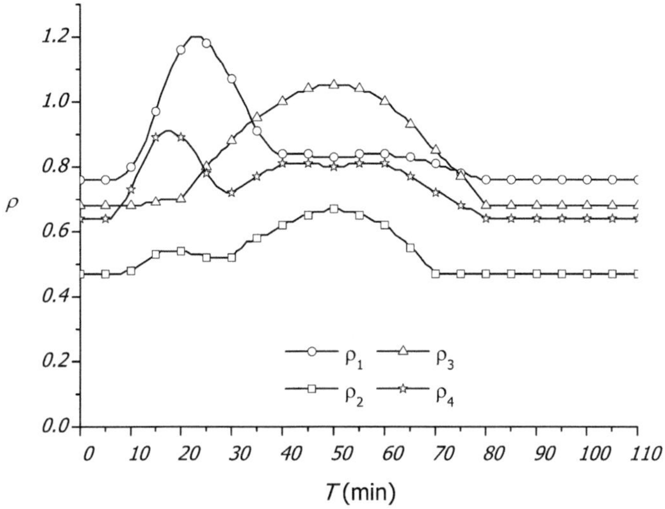

Fig. 4.5 Traffic intensity (degree of saturation) versus time at the four legs of a roundabout

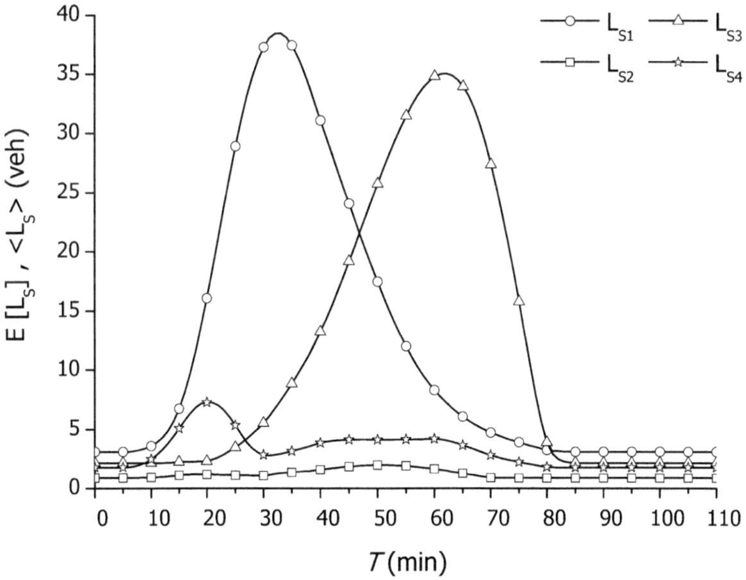

Fig. 4.6 The average number of users in the system versus time at the four legs of a roundabout

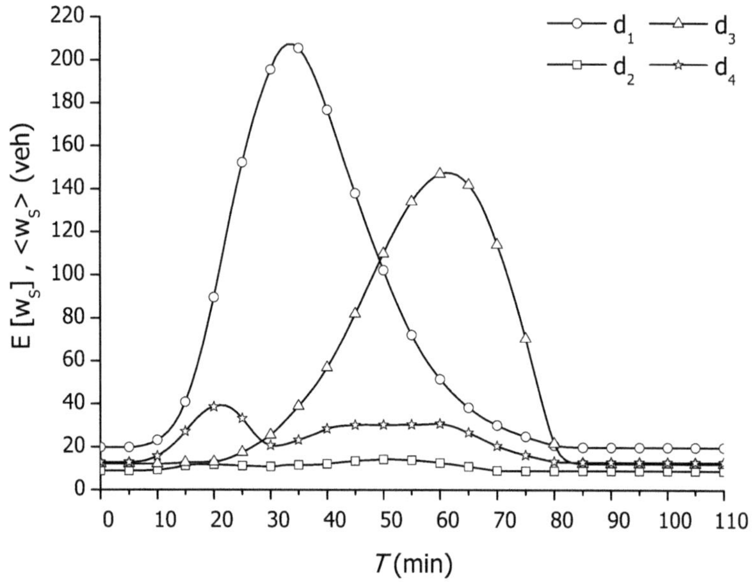

Fig. 4.7 The average time spent in the system versus time at the four legs of a roundabout

4.3 Evaluation of Levels of Service

For the evaluation of Levels of Service (LOS) of at-grade unsignalized intersections, we generally follow the indications of the American *Highway Capacity Manual* (HCM) [2].

The Manual distinguishes six LOS, from A to F, based on quality levels of circulation at the intersection, in descending order (A excellent; F very poor).

Among the variations introduced in the successive editions of the HCM, there is also a different selection of the parameters relative to Levels of Service.

In the edition published in 1985, the statement of LOS is given on the basis of intervals of reserve capacity RC = $C_i - Q_{ei}$ (See Table 4.1), and it gives only qualitative indications about delays.

Table 4.1 LOS, reserve capacity, and average delay according to HCM 1985

LOS	Reserve capacity	Average delay
A	>400	zero or short
B	300–400	short
C	200–300	medium
D	100–200	high
E	0–100	very high
F	<0	–

Table 4.2 LOS and average time in the system according to HCM 1994–HCM 1997 and HCM 2000

	HCM 94–HCM 97	HCM 2000	
LOS	Average delay (s)	Average delay (s)	Average time spent in the system (s)
A	<5	<10	<5
B	5–10	10–15	5–10
C	10–20	15–25	10–20
D	20–30	25–35	20–30
E	30–40	35–50	30–40
F	>45	>50	>45

In the editions published in 1994, 1997, and 2000, the Levels of Service are based on the levels of acceptability of the waits by the users, according to the values shown in Table 4.2.

Regarding the parameters shown in Table 4.2, we recall that "delay" means the time w defined with Eq. (1.21) of Chap. 1.

In Table 4.2, according to HCM 2000, it is assumed to be systematically greater than 5 s of the time spent in the system w_S.

It is worth noting that the classifications of LOS on the basis of reserve capacity RC or the waiting time are the same.

In fact, it is possible to demonstrate [3] that if we have a calculation procedure of the time spent in the system, we can identify a bijective relationship between the two parameters assumed to represent the Level of Service, and we can use that relationship to deduce the value of one of the parameters after the value of the other parameter is known.

Thus, for example, if the system is under steady-state conditions with Poissonian arrivals and exponential waiting times, from Eqs. (3.17) and (3.3) of Chap. 3, we immediately have,

$$E[w_S] = \frac{1}{C \cdot (1 - \rho)} = \frac{1}{C - Q_e} = \frac{1}{RC} \tag{4.58}$$

With Eq. (4.58), on the basis of the intervals of reserve capacity shown in Table 4.1, we have the intervals of waiting times shown in Table 4.3.

For RC < 0, Eq. (4.58) loses significance (See Fig. 3.3 of Chap. 3) consistently with the circumstance that, for RC $\leq$ 0, there are no statistical equilibrium conditions for the system.

Table 4.4 shows the correspondence calculated under steady-state conditions (RC > 0) in [4] between average times spent in the system $E[w_S]$ and reserve capacity, starting with various formulas for the evaluation of $E[w_S]$ and with reference to the intervals of reserve capacity shown in Table 4.1.

Table 4.3 LOS, reserve capacity, and average time spent in the system according to Eq. (4.58)

LOS	Reserve capacity	Average time spent in the system
A	<400	<9
B	300–400	9–12
C	200–300	12–18
D	100–200	18–36
E	0–100	>36
F	<0	–

Table 4.4 Reserve capacity and average time spent in the system according to various researchers

		Average time spent in the system (s)			
LOS	Reserve capacity	According to Kremser	According to Brilon and Grossman	According to Siegloch	According to Mauro
A	<400	<10	<10	<5	<10
B	300–400	10–12	10–15	5–9	10–15
C	200–300	12–15	15–25	9–15	15–22
D	100–200	15–20	25–45	15–30	22–44
E	0–100	>35	>45	>30	>44
F	<0	–	–	–	–

With the absence of steady-state conditions, because of saturation or oversaturation of the entries (RC $\geq$ 0) and/or because of the short duration of the period T_k of demand variation and/or capacity, a correspondence between reserve capacity and average times spent in the system may be obtained by expressing the time-dependent solutions as a function of RC instead of traffic intensity (degree of saturation) ρ.

Working in this way, it is possible to demonstrate that a reserve capacity RC greater than 100 veh/h is associated with a time spent in the system smaller than 40 s.

The threshold of 45 s given by the HCM 2000 for LOS E for at-grade unsignalized intersections (See Table 4.2) is reached for values of RC of 75–80 veh/h (See Fig. 4.8, [5]).

Anyway, under any condition of the system (steady-state or transient) in today's technical practice, in absence of specific alternative indications, we generally use the classification elaborated by HCM 2000 for linear, unsignalized intersections. According to this classification, the quality of the circulation at the intersection is established on the basis of the level of acceptability by the users of the waits at entries.

Specific studies of roundabouts have not been available. Therefore, the only solution is to use the criterion of HCM 2000 described in Table 4.2 for roundabouts.

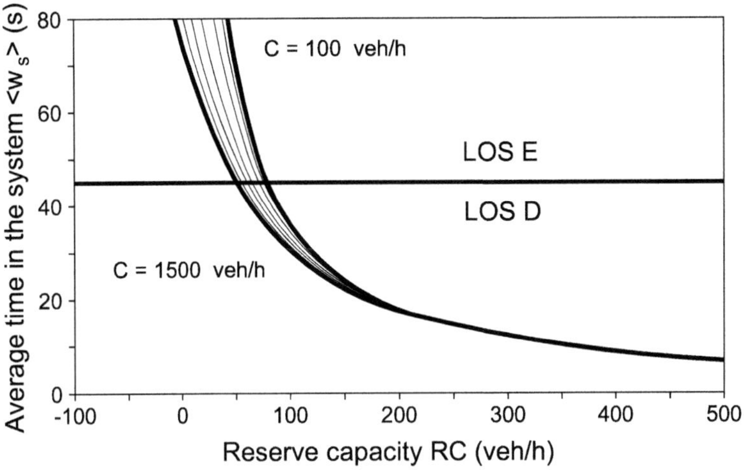

Fig. 4.8 Reserve capacity, time spent in the system, and levels of service [5]

The evaluation of LOS for roundabouts is performed by assuming that the roundabout entry that is characterized by the worst LOS determines the LOS of the entire intersection.

Finally, it is well known that short times are generally associated with transient conditions, so it is worth noting that the recording of the short times when there are poor Levels of Service does not invalidate the suitability evaluation of the intersection to perform its functions.

When this occurs, we must accurately evaluate the singular, actual situations on a case-to-case basis.

References

1. Brilon, W., Troutbeck, R., Tracz, M., *Review of International Practices Used to Evaluate Unsignalized Intersections*, Transportation Research Board Circular, number 468, Washington, April 1997.
2. Transportation Research Board, *Highway Capacity Manual*, Washington DC, edition 1985, 1994, 1997, 2000.
3. Brilon, W. (Editor), *Intersections Without Traffic Signals*, Vol. I, Springer-Verlag, Berlin, 1988.
4. Mauro, R., *Contributo alla progettazione delle intersezioni stradali non semaforizzate.* [Contribution to the design of unsignalized road intersections] (in Italian), Report of D.I.T. n°39/95, Naples, 1995.
5. Brilon, W., *Kreisel – Handbuch, Anhang A*, Version Oktober 2004.

Chapter 5
Evaluation of Roundabout Reliability

The study of the performances of roundabouts in terms of capacity indices (See Sect. 1.2) may be completed using further analysis in addition to the ones that we presented in the previous chapters.

This occurs because the flows of the various legs (and the capacities that depend on them) are random variables. In general, they are non-statistically independent. Therefore, to evaluate reliability, that is to say the probability that the system does not fail (in the specific case, that demand does not exceed the single entry capacity) it is necessary to characterize the flows and their related values by means of their probability functions or when these laws are not available, by coincise indices such as means, variances, and covariances.

This chapter presents general criteria for the evaluation of reliability in each leg based on the study of the performance function Z ($Z = C - Q_e$ or $Z = C/Q_e$) and it provides the analytical relations in the particular case in which capacity and demand (and thus also Z) are normally distributed with means and variances known.

An approximated criterion is also provided to be used in cases in which the probability laws of capacities and demands, and thus of performance function Z, are unknown or difficult to determine.

The worked examples developed to illustrate the method (and the many others that have not been reported for the sake of brevity) show, as it was logical to expect, that the two only indices normally used (Reserve Capacity and/or Rate of Capacity) are not sufficient to ensure that the system does not fail.

When the mean of the Reserve Capacity and/or of the Capacity Rate does not change, reliability depends on the level of uncertainty that affects the values in question (dispersion around the mean values of the flows).

Regarding the threshold value to attribute to reliability, it must be stated that it cannot be fixed in general terms, but it should be identified on a case-to case basis in relation to the damage (excessive mean and global waiting times, safety decrease, repercussions on the surrounding network) caused by the system failure.

The results presented in this chapter can help to analyze roundabouts in a better and more rational way.

Finally, even other performance indices for roundabouts, as simple and total capacity, can be expressed in a probabilistic way.

R. Mauro, *Calculation of Roundabouts*, DOI 10.1007/978-3-642-04551-6_5,

5.1 Reliability and Performance Functions

Consider a roundabout entry of capacity C, interested by an entering traffic demand Q_e.

To evaluate the reliability – intended as the value "Z" of the suitability of a system to perform its function – it is natural to compare the values C and Q_e.

The "Z" value, that is to say "performance function", can be measured both with the difference $C - Q_e$ and with the ratio C/Q_e, which in the reliability theory terminology [1] are indicated as reliability margin and reliability factor respectively:

$$Z = C - Q_e \tag{5.1}$$

$$Z = C/Q_e \tag{5.2}$$

Equations (5.1) and (5.2) are connected to the two most widely adopted roundabout capacity indices used to characterize service conditions: in fact, Eq. (5.1) coincides with the reserve capacity (RC), whereas Eq. (5.2) is the reciprocal of the rate of capacity when it is expressed in absolute terms (as we said in the previous Sect. 1.2).

Therefore, once prefixed two given minimum values z_{min} and z'_{min} for the reliability margin and reliability factor, the reliability condition of the system can be written:

$$Z = C - Q_e > z_{min} \tag{5.3}$$

$$Z = C/Q_e > z'_{min} \tag{5.4}$$

Particularly, in the former value the limit $z_{min}= 0$ can be taken and in the latter $z'_{min}=1$, so that the success condition is represented by

$$C - Q_e > 0 \tag{5.5}$$

$$C/Q_e > 1 \tag{5.6}$$

The complementary relations of Eq. (5.5) and Eq. (5.6) represent failure conditions:

$$C - Q_e < 0 \tag{5.7}$$

$$C/Q_e < 1 \tag{5.8}$$

What has been expounded so far is the deterministic position of the problem.

Because of the random nature of the factors and relations on which traffic capacity and traffic demand values depend, the previous values C and Q_e can vary randomly.

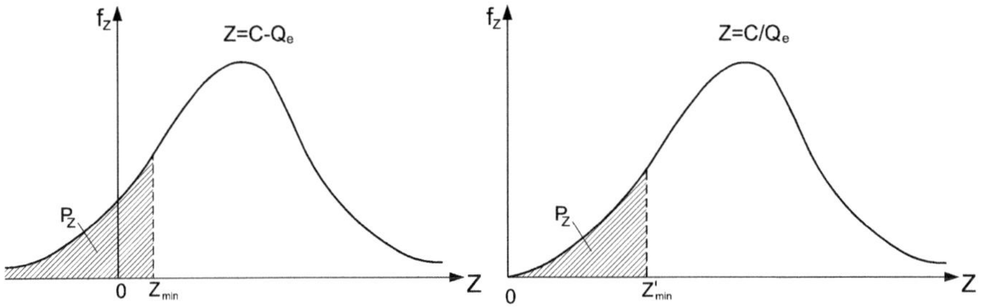

Fig. 5.1 Probability density function (p.d.f.) examples of the performance function (random variables Z)

Therefore, we consider C and Q_e as random variables described by given probability laws or by adequate statistics, the above-mentioned relations must be suitably modified to take this circumstance into account.

Thus, the values "Z" that are yielded by Eq. (5.1) and by Eq. (5.2) are also to be considered, insofar as they are random variable functions, as random values.

Also, it should be considered that each entering traffic demand value Q_e at an entry of capacity C corresponds to a reliability value determination.

Assume, then, in the more general case, that the probability law of the random variable "Z" (see Eqs. (5.1) and (5.2)) can be represented by the probability density function, p.d.f., f_Z (See Fig. 5.1).

On the basis of P_Z probability, the fractiles z_{min} and z'_{min} can be determined: this equals to let

$$P_Z = P\{Z = C - Q_e \leq z_{min}\} = F_Z(z_{min}) \tag{5.9}$$

$$P_Z = P\left\{Z = C/Q_e \leq z'_{min}\right\} = F_Z(z'_{min}) \tag{5.10}$$

where $F_Z(Z)$ are the distribution functions (c.d.f.) of the random variable "Z" in these two cases, $Z = C - Q_e$ and $Z = C/Q_e$.

P_Z is then the probability of the event {entry unreliability}, given that z_{min} and z'_{min} are minimum values prefixed by the reliability value.

The complementary event probability {entry reliability} obviously results in

$$1 - P_Z = P\{Z = C - Q_e > z_{min}\} = 1 - F_Z(z_{min}) \tag{5.11}$$

$$1 - P_Z = P\left\{Z = C/Q_e > z'_{min}\right\} = 1 - F_Z(z'_{min}) \tag{5.12}$$

The "failure" event and the complementary "success" event probabilities are then obtained respectively (See Fig. 5.2):

$$\left.\begin{aligned} P_f &= P\{Z = C - Q_e \leq 0\} = F_Z(0) \\ P_f &= P\{Z = C/Q_e \leq 1\} = F_Z(1) \end{aligned}\right\} \text{failure} \tag{5.13}$$

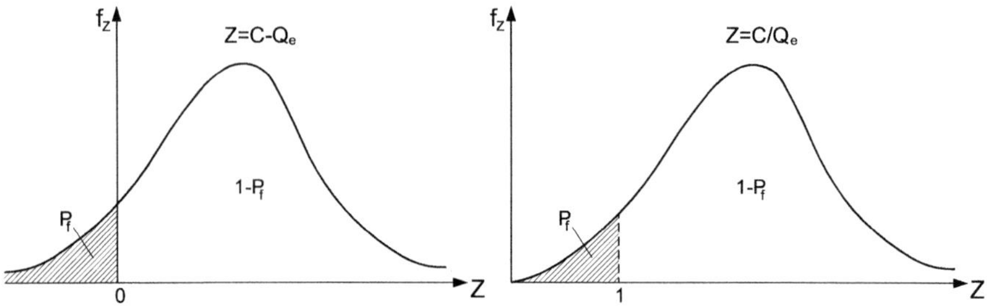

Fig. 5.2 Examples of the identification of the probability of "failure" and "success" events

$$\left.\begin{array}{l}\mathcal{A} = 1 - P_f = P\{Z = C - Q_e > 0\} = 1 - F_Z(0) \\ \mathcal{A} = 1 - P_f = P\{Z = C/Q_e > 1\} = 1 - F_Z(1)\end{array}\right\}\text{success} \tag{5.14}$$

From now on, the probability $1 - P_f$ associated with the "success" event will be briefly indicated by reliability $\mathcal{A}$. In particular, with reference to the performance variable $Z = C - Q_e$, if $f_{CQe}(c,q_e)$ is the combined p.d.f of C and Q_e, for Eq. (5.13) it results that:

$$P_f = P(Z \leq 0) = \int_D f_{CQ_e}(c,q_e)dc\,dq_e \tag{5.15}$$

In Eq. (5.15) D is the unsafe region, where the performance function $Z = C - Q_e$ takes values $Z \leq 0$ (See Fig. 5.3 for $(C, Q_e) \geq 0$). In other words, according to Eq. (5.15), the volume subtended by $f_{CQe}(c,q_e)$ in correspondence with the region D gives the value P_f.

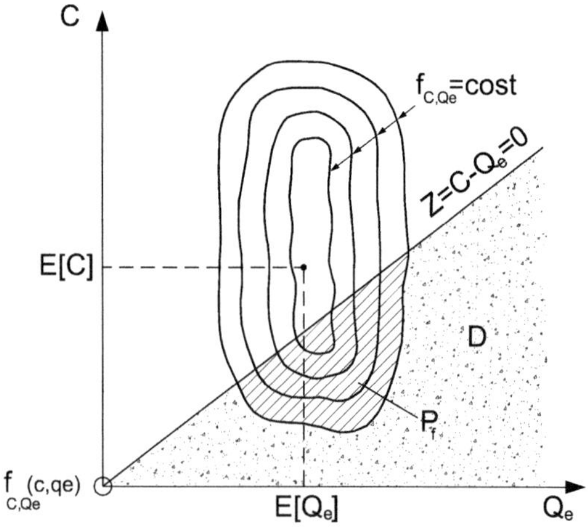

Fig. 5.3 Graphic identification of the unsafe region

In conclusion, the problem of the evaluation of a roundabout reliability refers, in its general formulation, to the thorough probabilistic characterization of the values that the entry flows Q_e can take and to the entry capacity C.

However, as it will be demonstrated in the following discussion, when this thorough characterization is not available, Level 1 reliability methods can be performed on the basis of statistical estimations of the expected value and of the standard deviation of the random variables Q_e and C, that is to say using only one measurement of the random variability of the values in question (the standard deviations $s_{Q_e} = \sqrt{VAR[Q_e]}$ and $s_C = \sqrt{VAR[C]}$, once the mean values $E[Q_e]$ and $E[C]$ are known).

5.2 General Calculation Procedure of Reliability

Adopting Eq. (5.1), $Z = C - Q_e$ as a performance variable for a common entry, if C and Q_e are random variables that are statistically independent, with the calculation rules for double integrals from Eq. (5.15) the two following equivalent expressions are obtained for P_f

$$\begin{aligned} P_f &= \int_0^{+\infty} f_C(c)[1 - F_{Qe}(c)]dc \\ P_f &= \int_0^{+\infty} f_{Qe}(q)F_C(q)dq \end{aligned} \tag{5.16}$$

where $f_C(\cdot)$ and $f_{Qe}(\cdot)$ are the p.d.f. respectively of C and Q_e, and $F_c(\cdot)$ and $F_{Qe}(\cdot)$ are the c.d.f. respectively of C and Q_e.

To prove Eq. (5.16) we start from Figs. 5.4 and 5.5.

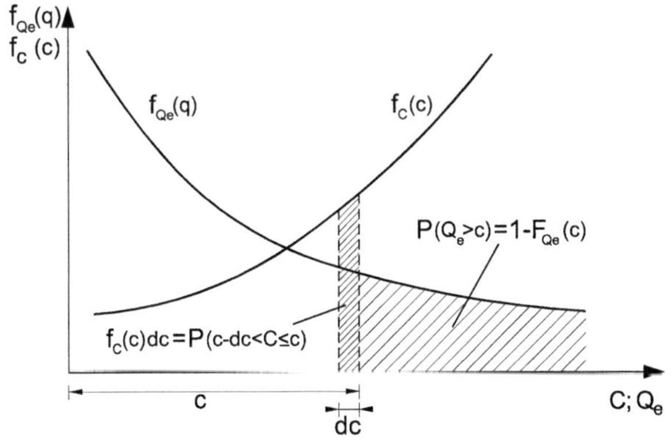

Fig. 5.4 Tails of p.d.f. of demand Q_e and of capacity C (generic p.d.f.)

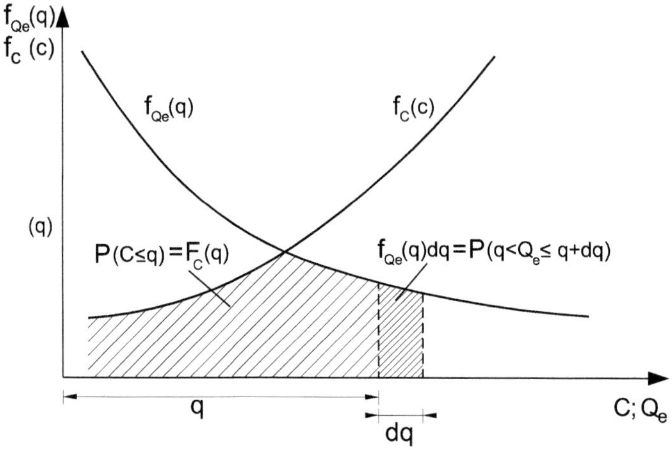

Fig. 5.5 Tails of p.d.f. of demand Q_e and of capacity C (generic p.d.f.)

Figure 5.4 shows the tails of the p.d.f. of demand Q_e and of capacity C (generic p.d.f.).

The probability that demand Q_e is bigger than an assigned value c of Capacity C is equal to

$$P(Q_e > c) = 1 - F_{Q_e}(c)$$

The probability that capacity C falls in the neighborhood of c is equal to

$$P(c - dc < C \leq c) = f_c(c)dc$$

The "failure" probability when C is equal to a given c is

$$dP_{f_c} = [1 - F_{Qe}(c)] \cdot f_c(c)dc$$

For all the possible c, it results

$$P_f = \int_0^{+\infty} [1 - F_{Qe}(c)] \cdot f_c(c)dc$$

which is Eq. (5.16).

Dually (See Fig. 5.5)

$$P(C \leq q) = F_c(q)$$
$$P(q < Q_e \leq q + dq) = f_{Q_e}(q)dq$$
$$dP_{f_{Qe}} = f_{Q_e}(q)F_c(q)dq$$

For all the possible q it results

$$P_f = \int_0^{+\infty} f_{Q_e}(q) F_c(q) dq$$

which is Eq. (5.16).

From Eq. (5.16), if C and Q_e are normally distributed, for P_f it results [1]

$$P_f = \frac{1}{\sqrt{2\pi}} \int_{-\infty}^{-\beta} e^{-t^2/2} dt = 0.5 - \text{erf}(\beta) \tag{5.17}$$

where erf(·) is the error function and β is the safety index (inverse of the coefficient of variation of the performance function Z)

$$\beta = E[Z]/\sqrt{VAR[Z]} \tag{5.18}$$

with E[Z] and $\sqrt{VAR\,[Z]}$ as the expected value and the standard deviation of the performance function Z respectively (Eq. (5.1)).

If C is normally distributed with mean $\overline{C}$ and variance σ_C^2 and Q_e is distributed, for example, exponentially with parameter α, for P_f it results that

$$P_f = F\left(-\frac{\overline{C}}{\sigma_c}\right) + \exp\{\,0.5\alpha^2\sigma_c^2 - \alpha\overline{C}\}\left[1 - F\left(-\frac{\overline{C}}{\sigma_c} + \alpha\sigma_c\right)\right] \tag{5.19}$$

where F(·) is the c.d.f. of the standardized normal distribution.

5.2.1 A Worked Example

For this example, the capacity formulation of SETRA [2] is adopted.

In Fig. 5.6 a roundabout with flow indications and geometrical features for the calculation of C capacity is schematically reported.

For all legs we have:

- entry width ENT = 4.00 m
- splitter island SEP = 6.00 m

The circulatory roadway width is ANN = 8.00 m.

For an entry capacity "i" the formulation selected gives

$$C = (1330 - 0.7 \cdot Q_{di}) \cdot [1 + 0.1 \cdot (ENT - 3.50)] \tag{5.20}$$

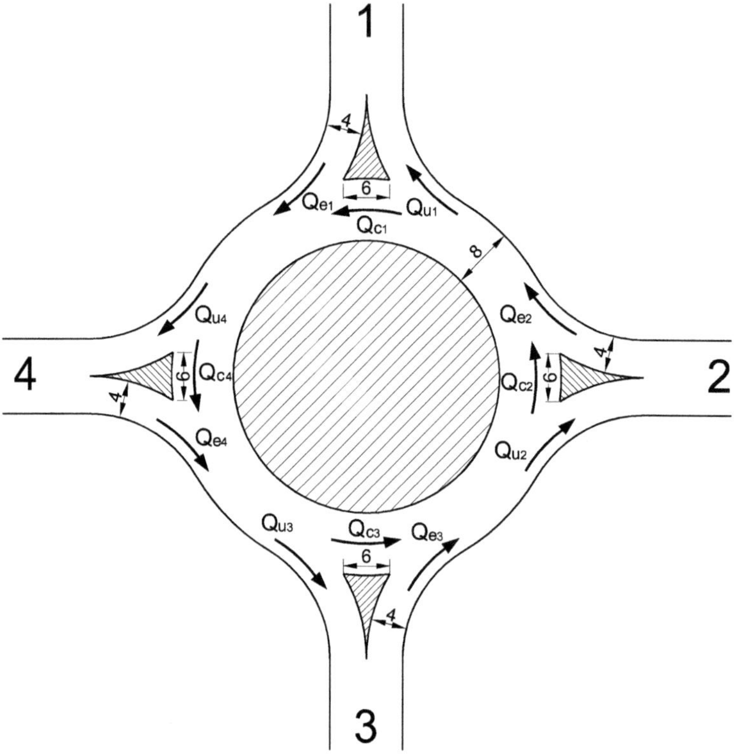

Fig. 5.6 Geometrical layout of the roundabout

where the disturbing flow Q_{di} is

$$Q_{di} = \left(Q_{ci} + \frac{2}{3}Q'_{ui}\right)[1 - 0.085 \cdot (ANN - 8)] \tag{5.21}$$

Q_{ci} is the circulating flow in front of the leg considered, Q_{ui} is the traffic exiting the leg i selected

$$Q'_{ui} = Q_{ui}\frac{15 - SEP}{15} \tag{5.22}$$

With the given values of ENT, SEP and ANN, Eqs. (5.20), (5.21) and (5.22) yield

$$C = 1397 - 0.735 \cdot Q_{di} \tag{5.23}$$

with

$$Q_{di} = Q_{ci} + 0.4 \cdot Q_{ui} \tag{5.24}$$

Table 5.1 Matrix O/D of the mean values $E[Q_{ij}]$ in (pcu/h) and variance $VAR[Q_{ij}]$ in $(pcu/h)^2$

O \ D	1	2	3	4	Q_{ei}
1	–	200	150	250	600
	–	2500	2000	2200	6700
2	300	–	100	150	550
	6400	–	1800	900	9100
3	300	200	–	100	600
	3000	2500	–	1800	7300
4	150	350	200	–	700
	1500	9000	2500	–	13000
Q_{ui}	750	750	450	500	–
	10900	14000	6300	4900	–

Supposing that, on the basis of experimental observations, mean flow $E[Q_{ij}]$ and variance $VAR[Q_{ij}]$ of all the turnings have been estimated and that these flows result in being statistically independent.

These data are reported in the matrix O/D of Table 5.1. In Table 5.1 in each square ij the top value is the mean, and the bottom one is the variance associated with the turning from leg i into leg j. Table 5.1 also shows the estimated values of the mean and variance for the overall flows for each leg at entry Q_{ei} and at exit Q_{ui}. The flows are expressed in pcu/h.

Mean and Variance of the Circulating Flows Q_{ci}

The circulating flows Qc are linear functions of traffic demand expressed as turning flows. For example, for entry 1 it results in $Q_{c1} = Q_{24} + Q_{23} + Q_{34}$.

On the basis of the properties of mean $E[\cdot]$ and variance $VAR[\cdot]$ operators, when applied to random variables statistically independent and with the data of Table 5.1, it gives

$$E[Q_{c1}] = E[Q_{24}] + E[Q_{23}] + E[Q_{34}] = 150 + 100 + 100 = 350 \text{ pcu/h} \quad (5.25)$$

$$\begin{aligned} VAR[Q_{c1}] &= VAR[Q_{24}] + VAR[Q_{23}] + VAR[Q_{34}] = \\ &= 900 + 1800 + 1800 = 4500\ (\text{pcu/h})^2 \end{aligned} \quad (5.26)$$

The coefficient of variation $(cv)_1$ is

$$(cv)_1 = \sqrt{VAR[Q_{c1}]}/E[Q_{c1}] = 67/350 = 0.192 \quad (5.27)$$

Doing the same for the other entries, the values of the mean and variance for the circulating flows in Table 5.2 can be obtained.

Table 5.2 Statistics of circulating flows, disturbing flows, and entry capacities. Safety index and reliability index values

Entry	$E[Q_{ci}]$ (pcu/h) $VAR[Q_{ci}]$ $(pcu/h)^2$	$E[Q_{di}]$ (pcu/h) $VAR[Q_{di}]$ $(pcu/h)^2$	$E[C_i]$ (pcu/h) $VAR[C_i]$ $(pcu/h)^2$	Average reserve capacity $E[C]-E[Q_e]$ (pcu/h)	β_i	A
1	350 4500	650 6244	920 3373	320	3.19	0.999
2	550 6300	850 8540	772 4613	222	1.90	0.971
3	700 13000	880 14008	750 7567	150	1.23	0.897
4	450 6300	650 7084	919 7084	219	1.55	0.939

Mean and Variance of disturbing flows Q_{di}

For entry 1 Eq. (5.24) gives $Q_{d1} = Q_{c1} + 0.4 \cdot Q_{u1}$ where

$$Q_{u1} = Q_{21} + Q_{31} + Q_{41}.$$

On the basis of the above-mentioned properties of the mean and variance operators and with the values of Table 5.1, it results in:

$$\begin{aligned} E[Q_{d1}] &= E[Q_{c1}] + 0.4 \cdot (E[Q_{21}] + E[Q_{31}] + E[Q_{41}]) = \\ &= 350 + 0.4 \cdot (300 + 300 + 150) = 650\,\text{pcu/h} \end{aligned} \tag{5.28}$$

$$\begin{aligned} VAR[Q_{d1}] &= VAR[Q_{c1}] + \overline{0.4}^2 \cdot (VAR[Q_{21}] + VAR[Q_{31}] + VAR[Q_{41}]) = \\ &= 4500 + 0.16 \cdot (6400 + 3000 + 1500) = 6244\,(\text{pcu/h})^2 \end{aligned} \tag{5.29}$$

In this case, the coefficient of variation equals to

$$(cv)_1 = \sqrt{VAR[Q_{d1}]}/E[Q_{d1}] = 79/650 = 0.122 \tag{5.30}$$

Repeating the calculation for the other entries, the disturbing flow statistics reported in Table 5.2 can be determined.

Mean and Variance of Flow Capacities C_i

For entry 1, with Eq. (5.23), since the relation between C_i and Q_{di} is linear and with the above-mentioned values calculated for Q_{di} moments, it results that

$$\begin{aligned} E[C_1] &= 1397 - 0.735 E[Q_{d1}] = \\ &= 1397 - 0.735 \cdot 650 = 920\,\text{pcu/h} \end{aligned} \tag{5.31}$$

$$\mathrm{VAR}[C_1] = \overline{0.735}^2 \cdot 6244 = 3373\,(\mathrm{pcu/h})^2 \tag{5.32}$$

$$(\mathrm{cv})_1 = 58/920 = 0.063 \tag{5.33}$$

For the other entries, the capacity moments reported in Table 5.2 can be determined in the same way.

Reliability Calculation

For entry 1, the performance function (5.1) Z_1 is on average equal to

$$\mathrm{E}[Z_1] = \mathrm{E}[C_1] - \mathrm{E}[Q_{e1}] = 920 - 600 = 320\,\mathrm{pcu/h}$$
$$\mathrm{VAR}[Z_1] = \mathrm{VAR}[C_1] + \mathrm{VAR}[Q_{e1}] = 3373 + 6700 = 10073\,(\mathrm{pcu/h})^2$$

It follows that

$$\beta_1 = \frac{\mathrm{E}[Z_1]}{\sqrt{\mathrm{VAR}[Z_1]}} = \frac{320}{\sqrt{10073}} = 3.19$$

and thus

$$\mathrm{erf}(\beta_1) = \mathrm{erf}(3.19) = 0.499.$$

With Eq. (5.17) reliability $\mathcal{A}$ for entry 1 on the basis of Eq. (5.14) equals to

$$\mathcal{A} = 0.5 + \mathrm{erf}(3.19) = 0.999.$$

Table 5.1 shows the reliability values calculated for the remaining entries to the roundabout considered. The present example has been carried out using the hypotheses of mutual statistical independence among entering flows. If this circumstance does not occur, it is also necessary to take into consideration the covariance $\mathrm{cov}[(\cdot);(\cdot)]$ of statistically dependent flows as shown below.

For example, with reference to entry 1, suppose that the elaboration of experimental data traffic shows the mutual statistical dependence among the circulating flows Q_{24}; Q_{23}; Q_{34} and among Q_{21}; Q_{31}; Q_{41}, so to obtain the values of Table 5.3 for the covariances (the Q_{ij} flows are expressed in pcu/h).

Table 5.3 Values of the $\mathrm{cov}[(\cdot);(\cdot)]$ in $(\mathrm{pcu/h})^2$

Q_{ij} \ Q_{ij}	Q_{23}	Q_{31}	Q_{34}	Q_{41}
Q_{21}		3500		2000
Q_{31}				1800
Q_{23}			1800	
Q_{24}	1000		1100	

While $E[Q_{c1}]$ remains equal to Eq. (5.25), with the values of Tables 5.1 and 5.3, the $VAR[Q_{c1}]$ is, with respect to Eq. (5.26), thus modified:

$$\begin{aligned} VAR\left[Q_{c1}\right] &= VAR\left[Q_{24}\right] + VAR\left[Q_{23}\right] + VAR\left[Q_{34}\right] + \\ &\quad +2\left(\text{cov}\left[Q_{24};Q_{23}\right] + \text{cov}\left[Q_{24};Q_{34}\right] + \text{cov}\left[Q_{23};Q_{34}\right]\right) = \\ &= 900 + 1800 + 1800 + 2\cdot(1000 + 1100 + 1800) = 12300(\text{pcu/h})^2 \end{aligned} \tag{5.26'}$$

For the disturbing flow Q_{d1} the same value yielded by Eq. (5.28) is obtained for the mean $E[Q_{d1}]$, while for the calculation of $VAR[Q_{d1}]$ it is necessary to know the $\text{cov}[(Q_{c1});(Q_{u1})]$, as well as the covariances of Table 5.3.

Suppose that for $\text{cov}[(Q_{c1});(Q_{u1})]$ it results, on the basis of traffic measurement treatment, $\text{cov}[(Q_{c1});(Q_{u1})]=13000\ (\text{pcu/h})^2$. With the values of Tables 5.1 and 5.3, it is obtained:

$$\begin{aligned} VAR\left[Q_{u1}\right] &= VAR\left[Q_{21}\right] + VAR\left[Q_{31}\right] + VAR\left[Q_{41}\right] + \\ &\quad +2\left(\text{cov}\left[Q_{21};Q_{31}\right] + \text{cov}\left[Q_{21};Q_{41}\right] + \text{cov}\left[Q_{31};Q_{41}\right]\right) = \\ &= 6400 + 3000 + 1500 + 2\cdot(3500 + 2000 + 1800) = \\ &= 25500(\text{pcu/h})^2 \end{aligned} \tag{5.34}$$

$$\begin{aligned} VAR[Q_{d1}] &= VAR[Q_{c1}] + \overline{0.4}^2 \cdot VAR[Q_{u1}] + 2\cdot 0.4\cdot \text{cov}[Q_{c1};Q_{d1}] = \\ &= 12300 + 0.16\cdot 25500 + 2\cdot 0.4\cdot 13000 = 26780\ (\text{pcu/h})^2 \end{aligned} \tag{5.35}$$

With Eq. (5.35) and with Eq. (5.28), on the basis of Eq. (5.20), for entry 1 capacity it is obtained that:

$$E[C_1] = 1397 - 0.735E[Q_{d1}] = 1397 - 0.735\cdot 650 = 920\,\text{pcu/h} \tag{5.36}$$

$$VAR[C_1] = \overline{0.735}^2 \cdot VAR[Q_{d1}] = \overline{0.735}^2 \cdot 26780 = 14467\,(\text{pcu/h})^2 \tag{5.37}$$

In the end, with $E[Q_{e1}]=600$ pcu/h (See Table 5.1) for the performance function Z_1 it results:

$$E[Z_1] = E[C_1] - E[Q_{e1}] = 920 - 600 = 320\,\text{pcu/h} \tag{5.38}$$

$$VAR[Z_1] = VAR[C_1] + VAR[Q_1] = 14467 + 6700 = 21167\,(\text{pcu/h})^2 \tag{5.39}$$

Thus

$$\beta_1 = \frac{E[Z_1]}{\sqrt{VAR[Z_1]}} = \frac{320}{145.5} = 2.20 \tag{5.40}$$

and for reliability $\mathcal{A}$ it results (See Eqs. (5.14) and (5.17))

$$\mathcal{A} = 0.5 + \text{erf}(\beta) = 0.5 + \text{erf}(2.20) = 0.986. \tag{5.41}$$

5.3 Approximated Calculation Procedure of Reliability

This procedure is exclusively based on the knowledge of the mean value and on only one measurement of the random variability (generally, variance or standard deviation) of the values involved: it can thus be considered an approximated approach to evaluate reliability compared to the criterion illustrated in the previous section.

Bearing in mind the meaning and the definition of mean value and variance, this method could be called the two-moment method. The use of this approximated method requires the introduction of a safety index β which can be defined using the performance function Z. This index is the number of standard deviations that separate the mean value Z from the value $Z = 0$ which – by definition – corresponds to the failure limit.

This method is structured as follows.

Calculate the mean and variance statistics E[Z] and VAR[Z] of the performance variable (5.1)

$$Z = C - Q_e \tag{5.1}$$

starting from the known homologues of Q_e and of C ($E[Q_e]$; $VAR[Q_e]$; [E[C];VAR[C]) with the relation (See Fig. 5.7)

$$E[Z] - \beta s_Z = 0 \tag{5.42}$$

index β is calculated, and it provides the number of standard deviations $s_Z = \sqrt{VAR[Z]}$ of Z that separate the mean value E[Z] from the value Z=0, corresponding by definition to the limit that marks the failure condition ($Z = 0 \leftrightarrow C = Q_e \ \forall\ Q_e;C \neq 0$).

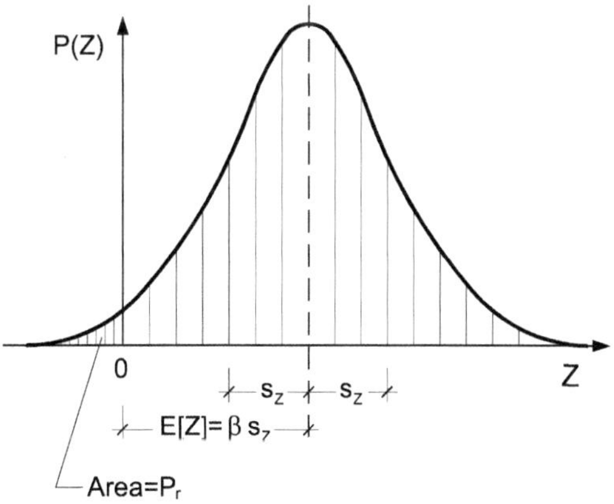

Fig. 5.7 Reliability index b and p.d.f. of the performance function Z

By introducing the normalized performance function as

$$\xi = \frac{Z - E[Z]}{s_Z} \tag{5.43}$$

the "success" and "failure" events are, respectively, equivalent to the occurrence of the inequalities:

– *success*

$$\xi \geq -\beta \tag{5.44}$$

– *failure*

$$\xi < -\beta \tag{5.45}$$

In fact, putting in Eq. (5.13) a value of $\xi \geq -\beta$ yields Z, $Z \geq 0$, that is to say that $C \geq Q_e$, while, substituting $\xi < -\beta$ yields $Z < 0$, which equals to $C < Q_e$.

If the Z probability law, that is ξ, is known, each limit of β corresponds to a well-determined value of the *failure probability*

$$P_f = P(\xi < -\beta = F(-\beta) \tag{5.46}$$

$$\mathcal{A} = 1 - P_f = 1 - F(-\beta) \tag{5.47}$$

Even though the law distribution of the performance function Z is not known or easily determinable, β can be considered as a coherent reliability value. In fact, the tail of the most common probability density functions can be adequately approximated with an exponential function (suitably identified with two parameters A and b, (A>0, b>0)),

$$F(\xi) = A \cdot \exp(b \cdot \xi) \quad \text{if} \quad F(\xi) =<< 1 \tag{5.48}$$

For example, if Q_e and C are both lognormally distributed, it can be demonstrated that Eq. (5.48) becomes

$$F(-\beta) = 460 \cdot \exp(-4.3\beta) \tag{5.49}$$

and, thus,

$$\mathcal{A} = 1 - F(-\beta) = 1 - 460 \cdot \exp(-4.3 \cdot \beta) \tag{5.50}$$

It follows that as β increases, reliability also increases. With the most frequent f.d.p. the values of β at least equal to 2 always indicate quite high probabilities that Q_e is systematically smaller than C, that is to say that the entry does not become saturated.

5.3.1 A Worked Example

Suppose we adopt as a capacity formulation the one presented in the 2000 edition of the HCM [3] on the basis of which

$$C = \frac{Q_c \exp\{-Q_c T_c/3600\}}{1 - \exp\{-Q_c T_f/3600\}} \text{ (pcu/h)} \tag{5.51}$$

For an assigned value of Q_c it results

$$C = C(T_c; T_f) \tag{5.52}$$

where T_c is the critical gap (s) and T_f is the follow-up time (s) (See Sect. 1.2).

We assume that Q_c is known without doubt and that T_c and T_f are instead random variables.

Using the linearization of Eq. (5.51) it can be demonstrated [4] that an approximated evaluation of the first order of E[C] and VAR[C] is yielded by

$$E[C] = C(E[T_c]; E[T_f]) = Q_c \cdot \frac{\exp\{-Q_c E[T_c]/3600\}}{1 - \exp\{-Q_c E[T_f]/3600\}} \text{ (pcu/h)} \tag{5.53}$$

$$VAR[C] = k_1^2 VAR[T_c] + k_2^2 VAR[T_f] + 2 \cdot k_1 \cdot k_2 \cdot cov[T_c; T_f] \tag{5.54}$$

where $E[T_c]$; $E[T_f]$; $VAR[T_c]$; $VAR[T_f]$; $cov[T_c;T_f]$ have the usual meaning and for k_1 and k_2 it results

$$k_1 = \left(\frac{\partial C}{\partial T_c}\right)_{E[T_c];E[T_f]} = -\frac{Q_c^2}{3600} \cdot \frac{\exp\{-Q_c E[T_c]/3600\}}{1 - \exp\{-Q_c E[T_f]/3600\}} \tag{5.55}$$

$$k_2 = \left(\frac{\partial C}{\partial T_f}\right)_{E[T_c];E[T_f]} = -\frac{Q_c^2}{3600} \cdot \frac{\exp\{-Q_c (E[T_c] - E[T_f])/3600\}}{(1 - \exp\{-Q_c E[T_f]/3600\})^2} \tag{5.56}$$

Assume therefore, for an entry for which $Q_c = 250$ pcu/h, the following values of the statistics of the random variables T_c and T_f:

$$E[T_c] = 4.4\,s \qquad VAR[T_c] = 0.36\,s^2 \qquad cov[T_c; T_f] = 0.30\,s^2$$

$$E[T_f] = 2.9\,s \qquad VAR[T_f] = 0.25\,s^2$$

With these values and with $C=C(T_c, T_f)$ given by Eq. (5.51), it is obtained for Eqs. (5.53), (5.54), (5.55) and (5.56)

$$k_1 = \left(\frac{\partial C}{\partial T_c}\right)_{E[T_c];E[T_f]} = -\frac{\overline{250}^2}{3600} \cdot \frac{\exp\{-250 \cdot 4.4/3600\}}{1 - \exp\{-250 \cdot 2.9/3600\}} = -70$$

$$k_2 = \left(\frac{\partial C}{\partial T_f}\right)_{E[T_c];E[T_f]} = -\frac{\overline{250}^2}{3600} \cdot \frac{\exp\{-250 \cdot (4.4 - 2.9)/3600\}}{(1 - \exp\{-250 \cdot 2.9/3600\})^2} = -470$$

$$E[C] = 250 \cdot \frac{\exp\{-250 \cdot 4.4/3600\}}{1 - \exp\{-250 \cdot 2.9/3600\}} = 1010 \text{ pcu/h}$$

$$VAR[C] = (-70)^2 \cdot 0.36 + (-470)^2 \cdot 0.25 + 2 \cdot 0.30 \cdot 32900 = 76729 \text{ (pcu/h)}^2$$

$\sqrt{VAR[C]} = 277$ pcu/h

The coefficient of variation (cv) is equal, in this case, to

$$(cv) = \sqrt{VAR[C]}/E[C] = 277/1010 = 0.27$$

If $E[Q_e] = 450$ pcu/h and $VAR[Q_c] = 1500$ (pcu/h)^2 it is obtained for the statistics of the performance function (5.1)

$$E[Z] = E[C] - E[Q_e] = 1010 - 450 = 560 \text{ pcu/h}$$

$$\sqrt{VAR[Z]} = \sqrt{VAR[C] + VAR[Q_e]} = \sqrt{76729 + 1500} = 280 \text{ pcu/h}$$

and thus, for β it results

$$\beta = \frac{560}{280} = 2.00$$

This value of β means high reliability values.

When it seems right to apply the approximation provided by Eq. (5.49) to this case the value (Eq. (5.50)) obtained for $\mathcal{A}$ is, for example,

$$\mathcal{A} = 1 - 460 \exp\{-4.3 \cdot 2.00\} = 0.92$$

5.4 Some Remarks

When the mean of the Reserve Capacity and/or of the Capacity Rate does not change, reliability depends, on the level of uncertainty that affects the values in question (dispersion around the mean values of the flows).

The role of the dispersions centered on the mean values $E[Q_e]$ and $E[C]$ is evident from the observation of the curves $P_f = P_f(\omega_o)$ in Fig. 5.8. They can be obtained as follows.

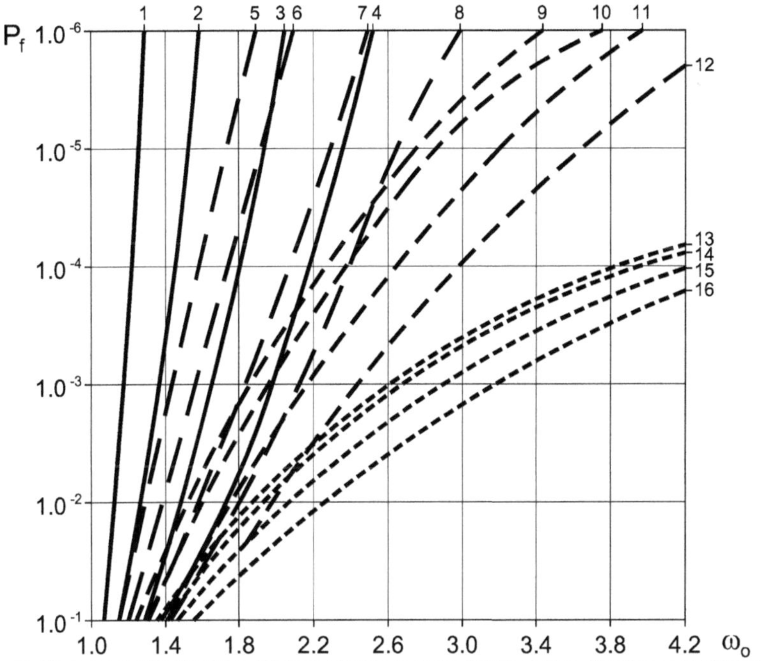

$(cv)_c$	$(cv)_{Qe}$			
	0	0.1	0.2	0.3
0.05	1	2	3	4
0.10	5	6	7	8
0.15	9	10	11	12
0.20	13	14	15	16

Fig. 5.8 $P_f = P_f(\omega_o)$ for the couples of values $((cv)_c; (cv)_{Qe})$

Expressed β as (Eq. (5.18))

$$\beta = \frac{E[C] - E[Q_e]}{\sqrt{VAR[C] + VAR[Q_e]}} = \frac{\omega_o - 1}{\sqrt{\omega_o^2 (cv)_c^2 + (cv)_{Qe}^2}}$$

where $(cv)_c = \sqrt{VAR[C]}/E[C]$ and $(cv)_{Qe} = \sqrt{VAR[Q_e]}/E[Q_e]$ are, respectively, the capacity and demand coefficients of variation at an entry and $\omega_o = E[C]/E[Q_e]$ the ratio among the means of the same ones, for Eq. (5.18) it results in

$$P_f = P_f(\omega_o; (cv)_c; (cv)_{Qe})$$

Once fixed the values that form the couples $((cv)_c; (cv)_{Qe})$ of the table, with them, from Eq. (5.17), the 16 curves $P_f = P_f(\omega_o)$ of Fig. 5.8 can be obtained.

Figure 5.8 shows that:

- for high values of $(cv)_c$, even increasing considerably ω_o, it is not possible to keep the failure probability within small values;
- for small values of $(cv)_c$, the variability of Q_e is significant (this is instead unimportant for big values of $(cv)_c$, that is to say with uncertain capacities).

References

1. Kottegoda, N., Rosso, R., *Statistic Probability and Reliability for Civil and Environmental Engineers*, MacGraw-Hill, New York, 1997.
2. SETRA/CSTR, *Capacité des carrefours giratoires interurbains,* Note d'information n°44, Paris, 1987.
3. *Highway Capacity Manual HCM 2000*, Special Report n. 209, T.R.B., Washington DC, 2000.
4. Benjamin, R., Cornell, C. A., *Probability, Statistics and Decision for Civil Engineers,* MacGraw-Hill, New York, 1970.

Index

Note: Locators in boldface refer to Definitions of entries.

R. Mauro, *Calculation of Roundabouts*, DOI 10.1007/978-3-642-04551-6,

GPSR Compliance

The European Union's (EU) General Product Safety Regulation (GPSR) is a set of rules that requires consumer products to be safe and our obligations to ensure this.

If you have any concerns about our products, you can contact us on ProductSafety@springernature.com

In case Publisher is established outside the EU, the EU authorized representative is:

Springer Nature Customer Service Center GmbH
Europaplatz 3
69115 Heidelberg, Germany

Batch number: 10614746

Printed by Printforce, the Netherlands